Interrelated Integrated Electronics Circuits FOR THE RADIO AMATEUR, TECHNICIAN, HOBBYIST, AND CB'ER

ROBERT M. MENDELSON
Applications Engineer
RCA Corporation, Solid State Division

HAYDEN BOOK COMPANY, INC.
Rochelle Park, New Jersey

ACKNOWLEDGMENT

While developing several of the circuits in this book, the author made use of the expertise of persons more versed in the fine points of certain integrated circuits. Harold Wittlinger and Robert D. Baird were most helpful, especially with operational amplifiers and discrete MOS devices. Special credit must be given to Mr. Baird for the design and layout of the Dwell-Tach circuit in Chap. 4.

Library of Congress Cataloging in Publication Data

Mendelson, Robert M
Interrelated integrated electronics circuits for the radio amateur, technician, hobbyist, and CB'er.

1. Integrated circuits—Amateurs' manuals.
I. Title.
TK9965.M46 621.381'73 79-18697
ISBN 0-8104-0760-4

Printed in the United States of America

1 2 3 4 5 6 7 8 9 PRINTING

79 80 81 82 83 84 85 86 87 YEAR

PREFACE

The purpose of this book is to provide a variety of projects that have appeal to the home builder. More than 25 useful circuits are described. They can be built by any Ham, technician, or hobbyist who is able to use a screwdriver, pliers, and soldering iron. Each circuit has been built, tested, and in many cases used to aid in the construction of later projects. Although the projects use solid-state integrated circuits, a few useful passive devices are included. Most projects are related to another project so that several may be combined into a single package for more versatility.

Circuit operation is discussed in everyday language. Construction details, layouts, and photographs are provided for each project to simplify its duplication. While most circuits are shown on printed circuit boards, they all can be hand wired using perforated boards and terminals. Sockets are used for all IC's to allow for easy replacement and servicing. However, direct soldering may also be used.

Room is left for the builder's originality as, for example, in the mechanical construction of the DVM and digital probes or in packaging some of the circuit boards to fit individual needs. This should broaden the home builder's construction abilities in the very interesting field of solid-state construction.

These are the general rules followed in all of the projects:

1. All resistors are 1/4 watt, 5% unless otherwise noted.
2. When 1% resistors are specified, they are necessary for good accuracy. Five percent resistors may be used, with a reduction in the overall circuit accuracy. With the aid of a good resistance decade box and Wheatstone bridge, selections of values may be made from 5% resistors.
3. Capacitors should have a voltage rating of greater than 25 volts. Capacitors with lower ratings are often leaky. When higher ratings are specified, there is a reason. Spikes or transients may occur. Any mica capacitor has a rating of several hundred volts.
4. While certain transformers, transistors, and integrated circuits are specified, equivalent components may be substituted.
5. Any circuit that is shown may be enclosed in a small box, if it so suits the user. Also, several circuits may be combined in one box for greater usefulness.

Some of the components in this book might prove difficult to obtain in a local electronics parts store. Specifically, some of the latest IC types that were used may not yet be available on the "surplus" market. They are, however, available for a reasonable price at any one of over 300 retail parts distributors that handle RCA components. Most of these stores can be located in the yellow pages of your telephone directory under "Electronic Equipment and Supplies" or "Electronic Parts." Almost any local parts store that displays the RCA logo stocks, or can order, RCA IC's or their SK series equivalent. As a last resort, contact RCA Distributor and Special Products Division, 2000 Clements Bridge Road, Deptford, New Jersey 08096, for information on distributor locations in your area.

ROBERT M. MENDELSON

Murray Hill, N.J.

Contents

1
Power Supplies and Accessories

Any electronic home construction project is incomplete without a power supply. Some systems require only one positive voltage source, while others operate better with a dual supply. For example, most operational amplifier circuits require both positive and negative voltages with a close 1:1 ratio between them. In this section several different power sources are described; both single and dual, low current and high current, as well as many useful accessories.

5- to 15-Volt Medium Current Power Supply

This first construction project will provide the hobbyist with a basic power supply that can be used as part of other projects or can be enclosed in a case and used as a small, well regulated bench supply. The power supply assembly is shown in Fig. 1-1.

Circuit Description

The schematic diagram of this circuit appears in Fig. 1-2. The secondary of the power transformer is connected to a full-wave rectifier. The 26.8-volt, center-tapped transformer T1 will provide close to 20 volts of filtered dc voltage at the input to the regulator. The entire regulation circuit is contained in one integrated circuit, the CA3085A(IC1). This device safely passes a maximum current of 100 milliamperes. Pins are provided for frequency compensation, short circuit protection, and current limiting.

The CA3085A has an internal reference voltage. In operation this voltage is continuously compared with a portion of the regulated output voltage, and the difference is fed to an error amplifier that corrects the output voltage.

Current limiting, and therefore short circuit protection, is provided by the voltage drop across R3 connected between pins 1 and 8 of the CA3085A. As the current through this resistor increases, a transistor in

Fig. 1-1 Medium current power supply finished assembly.

the IC slowly turns on. This reduces the total current available at the IC output. The resistance value of R3 determines the limiting current. If the CA3085A were to be used alone, the maximum current obtained would be 100 milliamperes. Since this is a rather small maximum current, an external series-pass power transistor Q1 has been added. The value of R3 was chosen to control the available base current to Q1 and, therefore, the maximum load current before limiting occurs. As shown, the circuit is capable of delivering approximately 300 milliamperes. The maximum current is also limited by the transformer chosen. A transformer with a secondary winding of 36 volts center tapped, such as is used in the tracking supply to be described later, would allow higher loads to be drawn.

The RC combination between pin 5 and ground provides frequency compensation and prevents oscillation.

Should it be desired to use this supply as part of another circuit where 100 milliamperes is adequate, the series-pass transistor may be eliminated. The output voltage would then appear at pin 8, which should be connected directly to the V+ terminal at the top of the divider. In addition, R3 should be decreased to 5.6 ohms. This will allow 100 milliamperes to be drawn before current limiting takes effect.

Finally, although this particular package was set to cover the range of 5 to 15 volts, there is no reason why a wider range (up to 35 volts) can't be covered. The maximum allowable input voltage to the CA3085A reg-

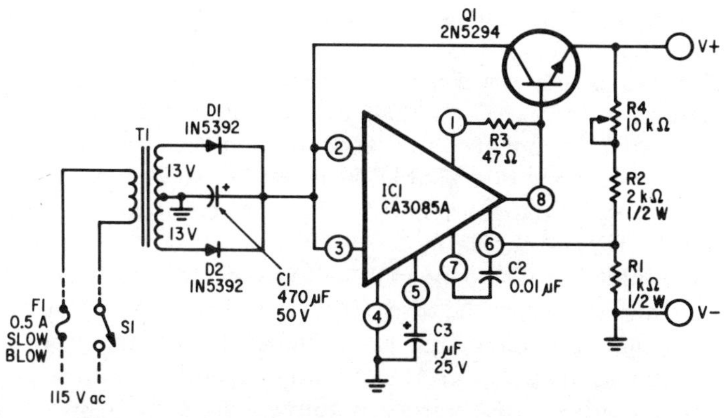

Fig. 1-2 Medium current power supply schematic diagram.

ulator is 40 volts, which will allow adjustment to 36 volts of regulated output. The dc input must be at least 4 volts higher than the desired output voltage for proper operation.

Do not use too high an input voltage if only low output voltages are required, because the difference between input and output must be dissipated by the series-pass transistor.

Parts List

C1—470-μF, 25-volt electrolytic
C2—0.01-μF, 100-volt ceramic
C3—1-μF, 25-volt electrolytic
D1, D2—1N5392 rectifier diode (also for D3 and D4 if required)
F1—0.5-ampere slow-blow fuse
IC1—RCA CA3085A voltage regulator
Q1—2N5294 npn power transistor
R1—1 kilohm, ½ watt
R2—2 kilohms, ½ watt
R3—47 ohms, ½ watt
R4—10-kilohm trim-pot
S1—SPST switch
T1—26-volt, 500-mA, center-tapped transformer (Triad F120-X or equivalent)
Misc.—Printed circuit board, IC socket, suitable enclosure, terminals, etc.

Construction

There are several extra holes in the printed circuit board. They were added to allow for modifications and to permit use of various components. Component locations are shown in Fig. 1-3.

If a center-tapped transformer cannot be obtained, any transformer with a secondary rating of greater than 13 volts may be used by adding two rectifier diodes to form a bridge rectifier. These are shown on the printed circuit board as D3 and D4. Be sure the diodes are inserted with the correct polarity configuration, as shown on the parts layout.

If an oscilloscope is available, it would be wise to check for oscillation over the entire voltage range. If the proper components are used, the circuit will be free of oscillation.

This printed circuit board may be installed in a small metal box for bench use. In such a case, the trim pot R4 could be removed and wires run to a panel potentiometer of the same value. Two binding posts will be required for attachment to external circuits. In addition, an ac switch and 0.5 ampere slow-blow fuse will be required.

While this is a single voltage supply, it may be used for plus or minus voltages by grounding the proper output terminal. Two of these supplies may be connected in series to provide a dual output. If this is done, care should be taken to see that the plus and minus supply voltages are adjusted individually to the proper values. (The third project is a tracking dual supply where both voltages are automatically kept equal. This is important in operational amplifier applications.)

With this supply, the hobbyist will be able to power many of the circuits in this book.

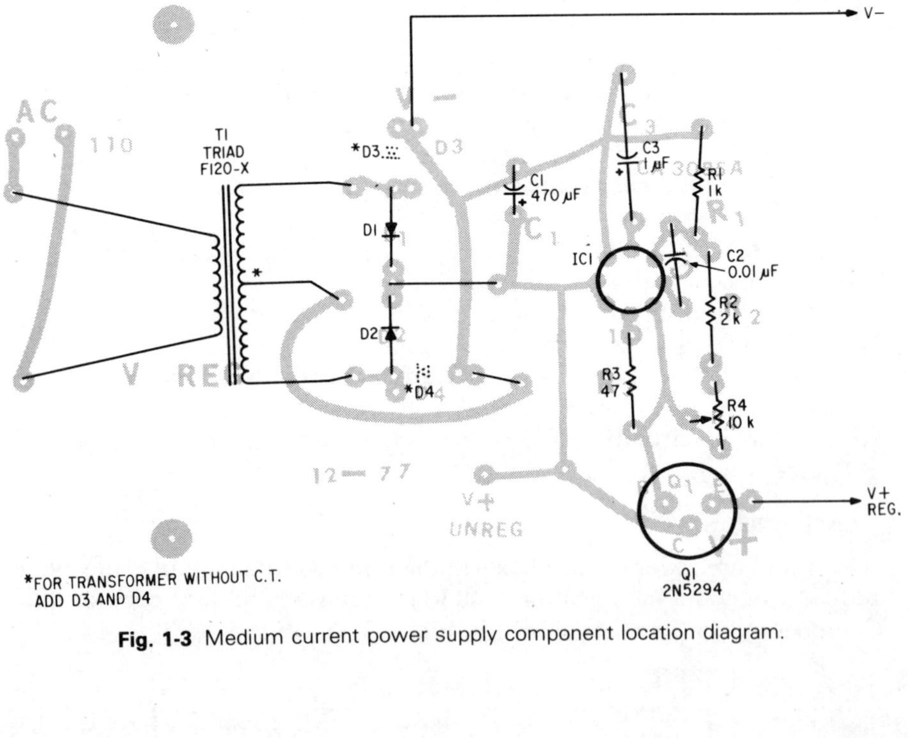

Fig. 1-3 Medium current power supply component location diagram.

5- to 15-Volt High Current Power Supply

This circuit is similar to the one previously described but it has several additions that allow for higher current with adequate short circuit protection. This supply can handle 5 to 15 volts at 2 amperes. Such power might be required to operate a low power Ham transmitter or receiver, an audio power amplifier, or any of dozens of experimental bench projects. Figure 1-4 shows the supply in an enclosure with the top removed.

Circuit Description

The schematic diagram of this circuit is shown in Fig. 1-5. The power transformer T1 delivers 17 ac volts each side of center. This is more than adequate for 15 volts output at full load. The CA3085A integrated circuit (IC1) operates the same as in the first power supply project. However, the series-pass transistor was chosen to pass the higher current without overheating. In addition, a series resistance of approximately 0.25 ohm (two 0.56-ohm resistors in parallel) has been added. Its purpose is the same at high current as R3 across pins 1 and 8 of the CA3085A was in the previously mentioned low current supply. When the total current reaches about 2.5 amperes, the increase in base drive to transistor Q2 reduces available output current from pin 8 of the CA3085A. This in turn reduces base drive

Fig. 1-4 High current power supply finished assembly with enclosure top removed.

of the series-pass power transistor Q1. This, therefore, provides current limiting and short circuit protection. In this circuit, the 6.8-ohm resistor from pin 1 to 8 is used to set the best base drive to Q1 and only partially

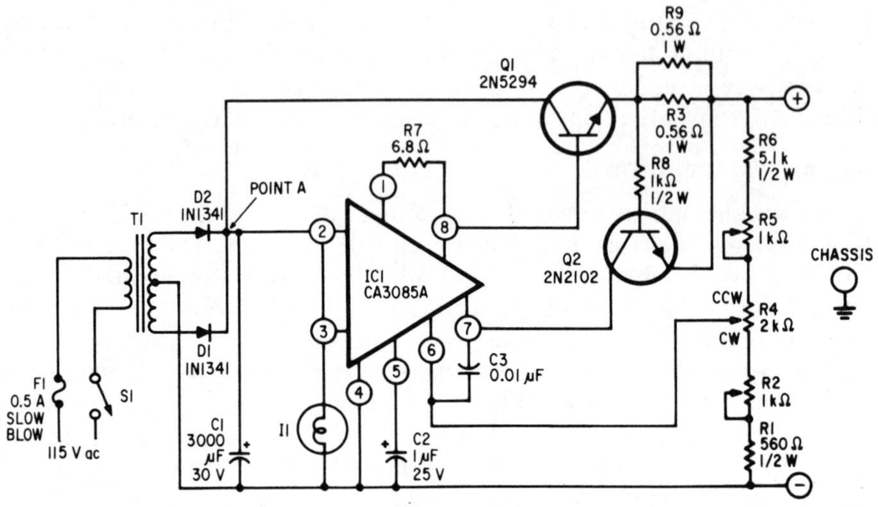

Fig. 1-5 High current power supply schematic diagram.

Parts List

C1—3000-μF, 30-volt electrolytic
C2—1-μF, 25-volt electrolytic
C3—0.01-μF, 100-volt ceramic
D1, D2—1N1341 rectifier diode
F1—0.5-ampere slow-blow fuse
I1—28-volt pilot light
IC1—RCA CA3085A voltage regulator
Q1—2N5294 npn transistor
Q2—2N2102 npn power transistor
R1—560 ohms, ½ watt
R2, R5—1-kilohm trim-pot
R3, R9—0.56 ohms, 1 watt
R4—2-kilohm potentiometer
R6—5.1 kilohms, ½ watt
R7—6.8 ohms
R8—1 kilohm, ½ watt
S1—SPST switch
T1—35-volt, outer-tapped transformer (Triax T54X or equivalent)
Misc.—Printed circuit board, IC socket, fuse holder, suitable enclosure, terminals, heat sink, etc.

affects overall current limiting. As was discussed in the first project, higher output voltages may be obtained by use of the proper transformer with up to 40 volts input to the regulator to give 36 volts out.

Construction

As an example of an enclosed bench supply, this printed circuit board was mounted in a small cabinet (see Fig. 1-4) together with all the extra components. Component location is shown in Fig. 1-6. The front

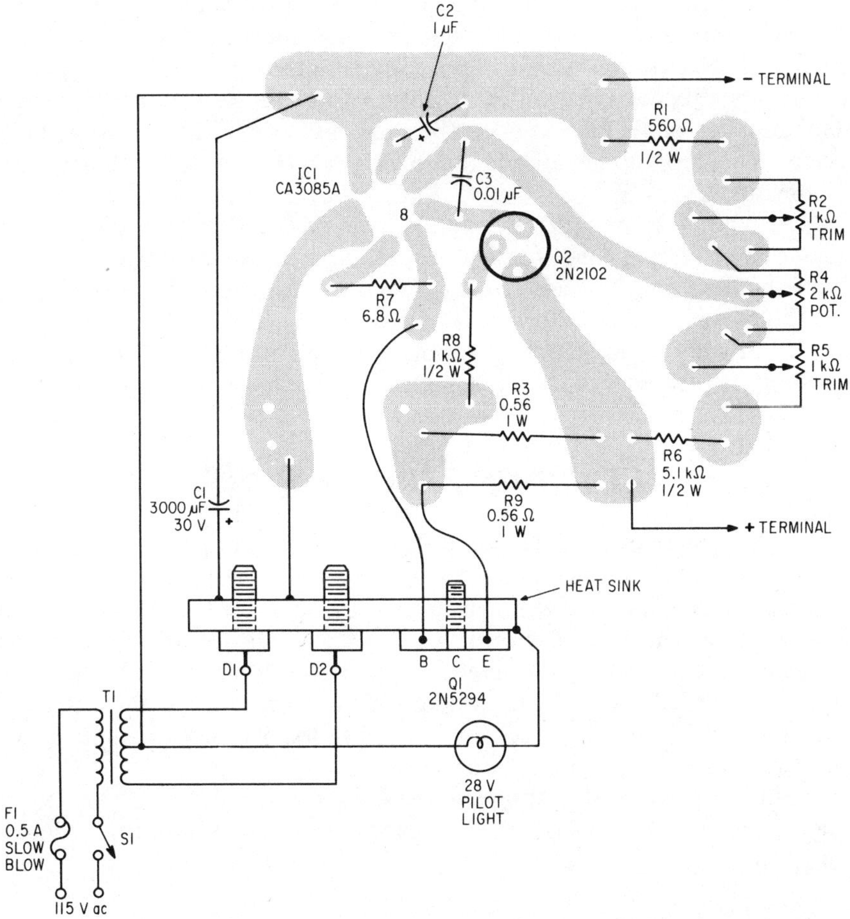

Fig. 1-6 High current power supply component location diagram.

panel is self-explanatory. The output voltage settings were hand calibrated around the voltage control knob.

The size of the cabinet prevented the addition of a permanent voltmeter, but one could be added if a larger box is used. The printed circuit board was made by use of a resist marking pen and etch solution. The main construction point to watch is that both rectifier diodes and the series-pass transistor Q1 are bolted directly to a heat sink with no insulating micas. A check of the schematic will show that point A is common to the cathodes of the diodes and the collector (case) of the power transistor. The only insulation required is a pair of plastic or porcelain threaded spacers to isolate the heat sink from the cabinet—this is important. Use an ohmmeter to be sure the heat sink is not shorted to the cabinet before the power is first applied.

To set the trimmers R2 and R5, attach a voltmeter to the output terminals. Turn control pot R4 fully clockwise and adjust R2 for a reading of 15 volts. Next, set the control pot fully counterclockwise and adjust R5 for 5 volts. Repeat once more or until a total range of 5 to 15 volts is obtained. The panel markings can be made by setting the control knob to give a measured voltage and then applying rub-on characters in appropriate positions.

As with the low current supply, two of these may be series connected to provide a dual voltage output. Although unwanted oscillation should not be present if the layout shown is used, it would be wise to make an oscilloscope check before putting the supply to use.

With this high current supply, power amplifiers, small Ham transmitter–receivers, and other power devices may be put to work.

Tracking Dual Supply

It has been previously mentioned that two supplies may be connected in series to provide positive and negative outputs with respect to a zero reference point between. Supplies like this are often needed for operational amplifiers. However, this method has the disadvantage of requiring independent adjustment to give a match between the plus and minus outputs. The circuit described in this project is self-tracking. That is, if the positive supply is set for plus 12 volts, the negative supply will automatically be set to minus 12 volts.

Since operational amplifiers usually operate in the 7- to 15-volt range, there is no need for higher voltages in this tracking supply, shown in Fig. 1-7.

Circuit Description

The CA3085A regulator circuit is similar to those used in the first and second projects. The positive output voltage is adjusted by R4 as before. The negative output voltage is regulated by the CA3094 operational power amplifier and the two 10-kilohm, 1% resistors R5 and R6 which are used to make a 1:1 ratio divider. Note that the ground terminal (see Fig. 1-8) is

Fig. 1-7 Tracking dual supply finished assembly.

connected to the noninverting input pin of the CA3094. The pair of 10-kilohm resistors are connected as a divider between the plus and minus output terminals with the center point connected to the inverting input of the CA3094. The operational amplifier compares the voltage of this point with the ground voltage at the noninverting input. The output voltage to the divider of the negative terminal is automatically adjusted to keep the inputs of the CA3094 equal.

The accuracy of the positive and negative match is directly related to the accuracy of the divider resistor match. This is why 1% resistors are used.

To allow for 500-milliampere capability, Q1 and Q2 have been added as series-pass transistors. Current limiting in the plus output is controlled by R3 and the internal circuit of the CA3085A.

The negative supply current limiting is provided by R7, R10, and Q3. This circuit operates in the same manner as the limiter in the 2 ampere supply of the second project except the collector and emitter leads of Q3 have been reversed.

Construction

The circuit is completely contained on one printed circuit board. Component location is shown in Fig. 1-9. The transformer may be any one that will supply 19 volts dc under load across each filter capacitor.

This supply has been enclosed in a small cabinet, shown in Fig. 1-10 with the top removed. A miniature voltmeter is included to make it easy to

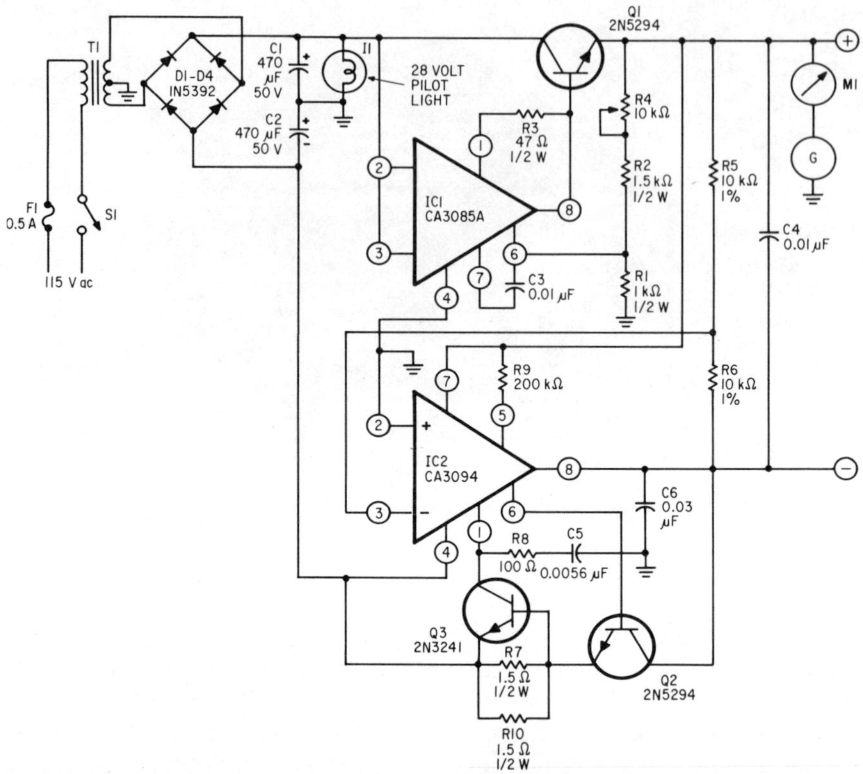

Fig. 1-8 Tracking dual supply schematic diagram.

Parts List

C1, C2—470-μF, 50-volt electrolytic
C3, C4—0.01-μF, 100-volt ceramic
C5—0.0056-μF, 80-volt tubular
C6—0.03-μF, 100-volt ceramic
D1 through D4—1N5392 rectifier diode
F1—0.5-ampere slow-blow fuse
I1—28-volt pilot light
IC1—RCA CA3085A voltage regulator
IC2—RCA CA3094 power op amp (or SK-3617)
M1—0-25-volt voltmeter
Q1, Q2—2N5294 npn power transistor
Q3—2N3241 small-signal transistor
R1—1 kilohm, ½ watt
R2—1.5 kilohms, ½ watt
R3—47 ohms, ½ watt

R4—10 kilohm trim-pot
R5, R6—10 kilohms, 1%
R7, R10—1.5 ohms, ½ watt
R8—100 ohms
R9—200 kilohms
S1—SPST switch
T1—36-volt, 1-ampere, center-tapped transformer (Stancor P8671 or equivalent)
Misc.—Printed circuit board, IC sockets, suitable enclosure, terminals, fuse holder, etc.

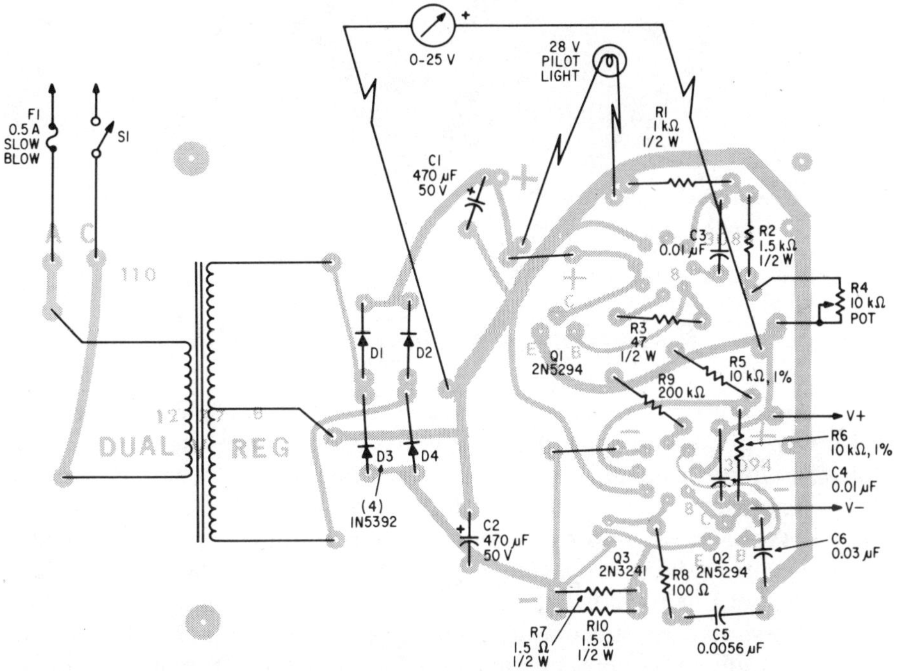

Fig. 1-9 Tracking dual supply component location diagram.

read the output voltages. Only the positive supply voltage is read since the negative supply will always match it.

As with the previously mentioned power supplies, use an oscilloscope to check for stray oscillations before putting the supply to use. With the layout shown, there should be no trouble.

With this tracking positive and negative supply, it will be much easier to utilize hundreds of operational amplifier circuits that appear in the literature.

Fig. 1-10 Inside view of the tracking dual supply.

Regulated Voltage Divider

As the hobbyist's circuits become more and more complex, he often discovers that two regulated supplies are needed to operate the circuit. A plus and minus tracking supply for use with op amps has been described.

Suppose, however, two plus supplies are needed. For example, plus 12 volts for a linear amplifier circuit and plus 5 volts for some TTL digital components in the same package. Two complete regulated supplies could be built, but this would require a duplication of every component. Alternately, a simple resistor divider could be used to tap the 5 volts from the 12-volt supply. This approach has two disadvantages. The main regulator would have to be capable of handling the total current load, and the regulation from the resistor tap would not be the best. An operational amplifier can be used with a minimum number of parts to provide a well regulated 5 volts without adding any load to the 12-volt regulator. Such a circuit is described here. The finished project appears in Fig. 1-11.

Circuit Description

Figure 1-12 shows the schematic diagram for this project. The CA3094 IC is a power op amp. By power, it is meant that the output stage has a power transistor capable of handling 100 milliamperes continuous with 300 milliamperes peak current. The schematic shows that the 12-volt

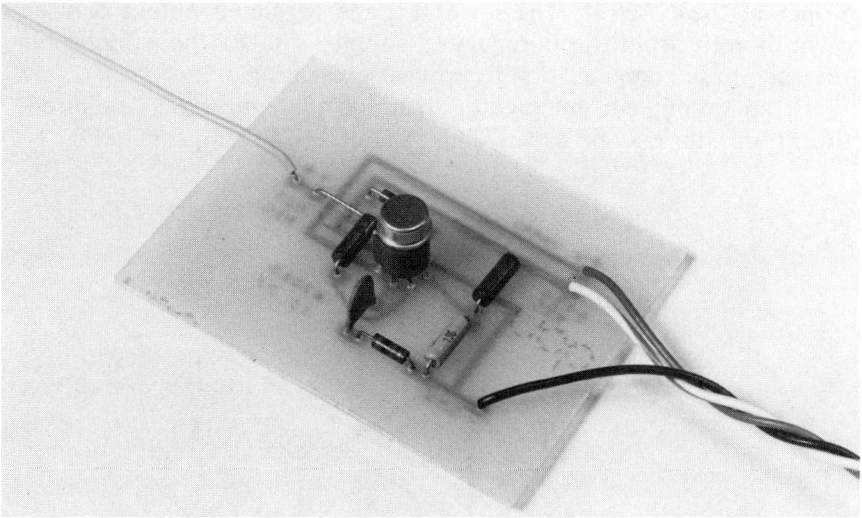

Fig. 1-11 Regulated voltage divider finished assembly.

regulated supply is used to operate the op amp, but the load current for the 5-volt supply comes directly from the rectified dc input voltage. This is possible because both emitter (pin 6) and collector (pin 8) of the power transistor are not connected internally but are free floating. The op amp is connected for unity gain. As a result, the output voltage at pin 6 will be equal to the voltage at the noninverting input of pin 3. This voltage is set by selection of R2 and the value of the regulated voltage available. The collector of the output power transistor is connected to the nonregulated dc power source.

Since the input current of the CA3094 is very small, it will have no effect on the 12-volt regulation; nor will its low operating current.

The output voltage can be calculated from the equation

$$V_{out} = \frac{R2}{10k + R2} \times V_{cc\ Reg}$$

where R2 is in kilohms.

In the example shown

$$V_{out} = \frac{7.3k}{10k + 7.2k} \times 12 = 5 \text{ volts}$$

A 7.3-kilohm value was chosen as the nearest standard 1% resistor.

Again, the advantage of this regulated divider lies in the use of the higher regulated voltage to only supply less than 1 milliampere of current to operate the CA3094. The lower voltage regulated output current is drawn directly from the unregulated supply. The RC network at pin 1 provides phase compensation to prevent oscillation.

If an output current greater than 100 milliamperes is required, a power transistor may be added. This modified circuit appears in Fig. 1-13.

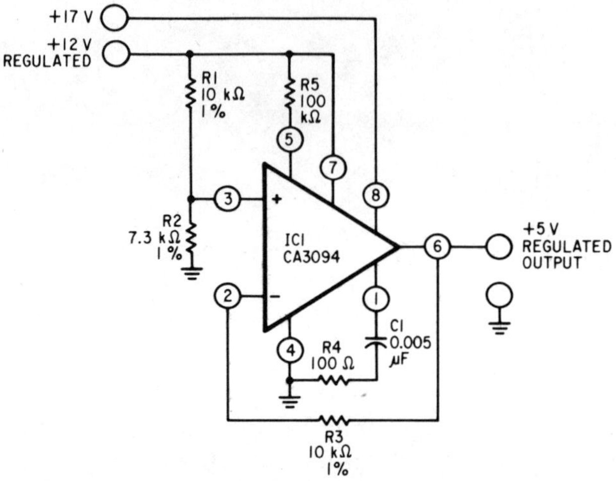

Fig. 1-12 Regulated voltage schematic diagram.

Parts List

C1—0.005-μF, 100-volt electrolytic
IC1—RCA CA3094 power op amp (or SK-3617)
R1, R3—10 kilohms, 1%
R2—7.3 kilohms, 1%
R4—100 ohms
R5—100 kilohms
Misc.—Printed circuit board, IC socket, terminals, suitable enclosure, etc.

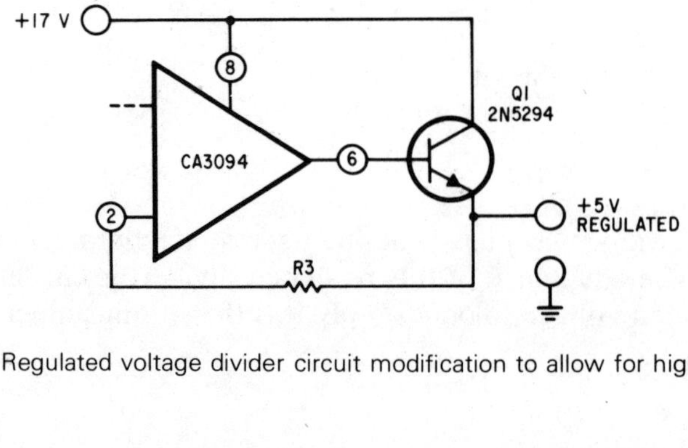

Fig. 1-13 Regulated voltage divider circuit modification to allow for higher current.

Construction

As in the previous circuits, there are no critical components here. The printed circuit layout is quite simple (Fig. 1-14) and may be added to a larger circuit if desired. Hand wiring is also quite acceptable.

When the circuit is complete, apply the regulated and unregulated input voltages and measure the output voltage. Fine tuning of this voltage may be made by adjustment of the value of R2.

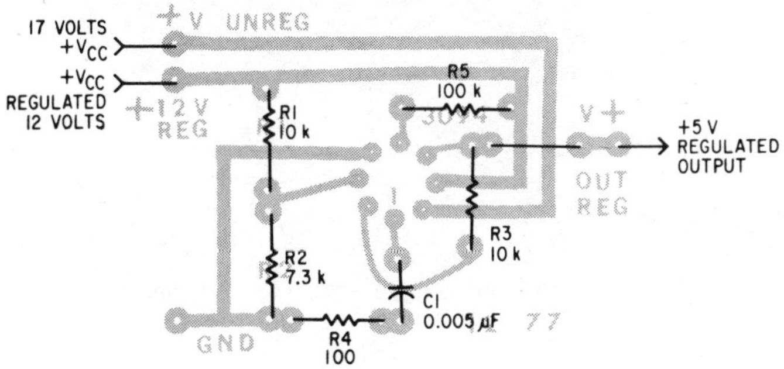

Fig. 1-14 Regulated voltage divider component location diagram.

A point of interest: If the low voltage load current is only a few milliamperes, so that it will not overload the regulated 12-volt supply, the collector of the power transistor (pin 8) may be connected to the +12 regulated supply instead of to the dc unregulated power source.

Thus, a simple operational amplifier circuit allows two different regulated voltages to be provided with a minimum of circuit parts.

Stable Voltage Reference

It is often necessary to have a stable fixed reference voltage in the design of a circuit. Circuits requiring a reference voltage include voltage controlled oscillators, phase-locked loop circuits, comparators, and regulated power supplies.

In addition, it is nice to have a standard voltage to check the calibration of voltmeters, or especially to set the amplitude of signals on an oscilloscope display. The circuit to be described is easy to build, costs little, is stable, and supplies a repeatable voltage from day to day. The finished project appears in Fig. 1-15.

Circuit Description

The complete schematic is shown in Fig. 1-16. The operational amplifier IC1 is running at full open-loop gain. The two inputs are used to compare the voltage between a reference diode and the tap on a resistance divider.

Fig. 1-15 Stable voltage reference finished assembly.

The Zener diode D1 sets a fixed voltage on the noninverting input of the operational amplifier. The inverting input is forced to this same voltage by the open-loop gain of the op amp and the resistance divider. The voltage on the output must rise to some higher voltage to allow for the voltage division by R1 and R2. The actual voltage at the op amp output can be changed by varying the ratio of the resistors in the divider. Note that the Zener diode is tied to this stable output voltage rather than the V_{cc} line to provide still greater stability. A decade voltage divider has been provided so that 10, 1, or 0.1 volts are available.

Construction

The entire circuit is mounted on a small printed circuit board. The board, in turn, is attached to the switch to make a compact, one-piece device. This circuit is simple enough to warrant hand wiring rather than printed circuit construction. Component location for the latter method appears in Fig. 1-17.

One percent resistors are used for increased accuracy and stability. When the wiring is complete, the input voltage can be applied. Any voltage from 12 to 15 volts will be fine. The voltage does not have to be regulated, but slightly better repeatability will be achieved if it is.

Use a VTVM or DVM to measure the reference output voltage with the switch set for 10 volts. If the reading is not close enough to 10 volts, a fine adjustment may be made by varying R1 or R2. If the output voltage is too high, add a high value resistor in parallel with R1. Start with 100 kilohms and reduce it till the best value is found.

If the output voltage is too low, add the shunt resistor across R2. The printed circuit board has provisions for these shunts. An inside view of the completed project is given in Fig. 1-18. Once this reference voltage is built, the hobbyist will find many uses for it and will wonder how he did without one for so long.

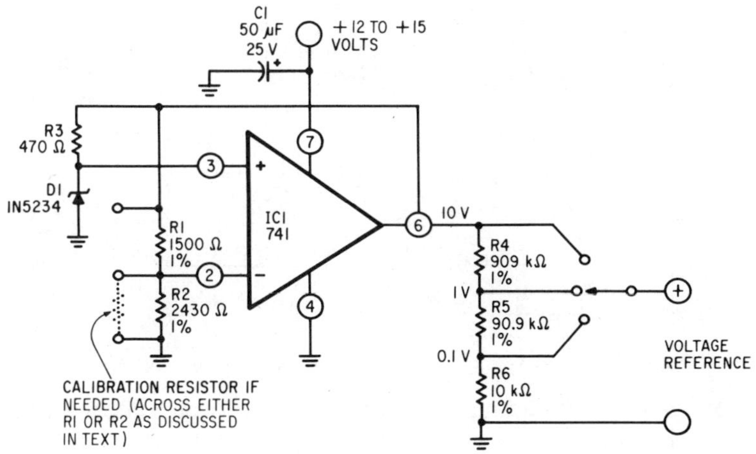

Fig. 1-16 Stable voltage reference schematic diagram.

Parts List

C1—50-μF, 25-volt electrolytic
D1—1N5234 Zener diode
IC1—741 op amp
R1—1.5 kilohms, 1%
R2—2.43 kilohms, 1%
R3—470 ohms
R4—909 kilohms, 1%
R5—90.9 kilohms, 1%
R6—10 kilohms, 1%
S1—Single-pole, three-position switch (Centralab PA-1001 or equivalent)
Misc.—Printed circuit board, suitable enclosure, terminals, IC socket, calibration resistor if needed, etc.

Constant Current Source

An amplifier whose output is a source of constant current regardless of the load has many uses. If an unknown value of resistance is inserted as the load, the voltage measured across the resistor is a direct function of the resistance. In other words, the DVM can be used as a direct reading

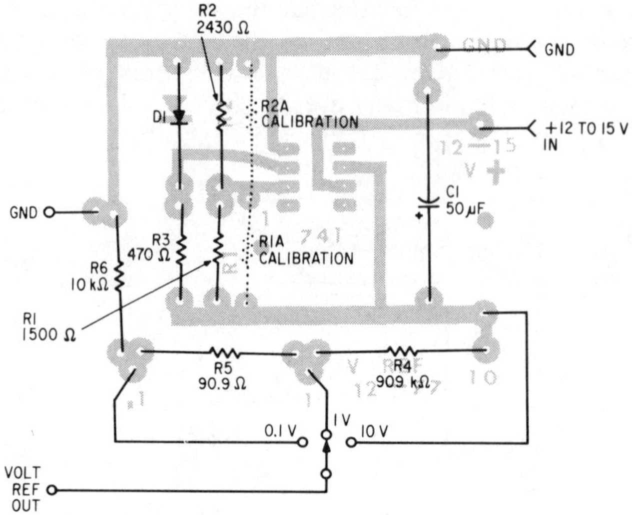

Fig. 1-17 Stable voltage reference component location diagram.

ohmmeter, thereby extending its usefulness. The circuit does not have to be mounted in an individual box as shown in Fig. 1-19. It can be incorporated into the DVM box with a switch to make the conversion from DVM to ohmmeter. In fact, this circuit can be used as part of any larger circuit where a constant current is needed.

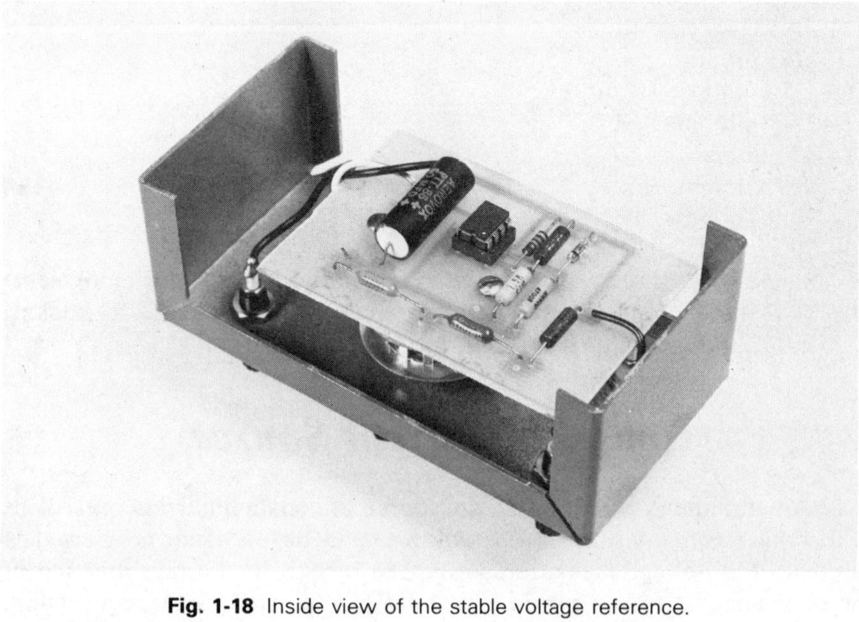

Fig. 1-18 Inside view of the stable voltage reference.

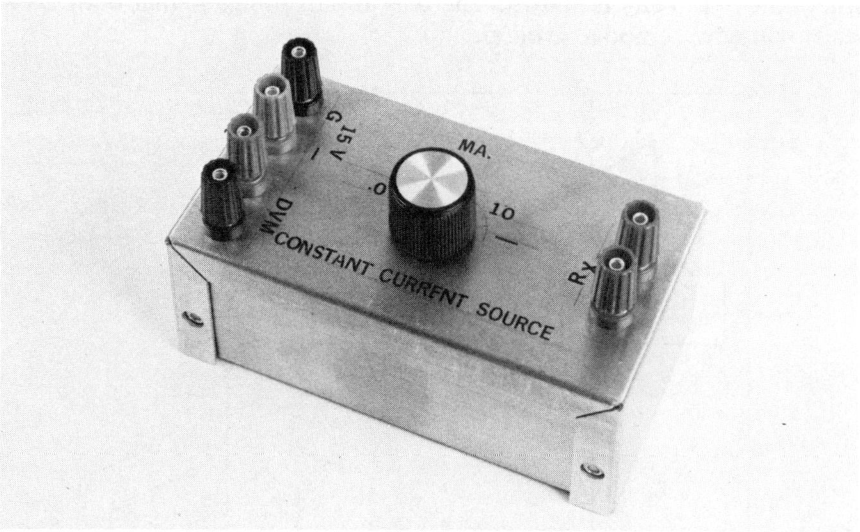

Fig. 1-19 Constant current source finished assembly.

Circuit Description

The CA3140 operational amplifier and the discrete MOSFET 40673 are connected as a unity gain amplifier (Fig. 1-20). The noninverting input is connected to a voltage divider at a point 10 volts above the −15 volt supply. In a unity gain amplifier, the output will assume the same voltage. This means that the voltage across the four switchable output resistors will also be 10 volts above the −15 volt supply.

Since both voltage and resistor are fixed, the current through the resistor must also be fixed. The unknown resistor is placed between the discrete MOSFET and the most positive point in the circuit. The fixed current into the MOSFET source must flow out through its drain and through the unknown resistor. The voltage measured across the resistor is therefore a direct function of the unknown resistance.

In this circuit, a switch allows the choice of four currents in multiples of 10 so that it may be used for a direct reading ohmmeter. One switch position is left unused. The builder may select any other current he might want.

The resistor value to be chosen for a given current can be calculated from

$$R = \frac{10 \text{ volts}}{I_{const}}$$

This circuit will work equally well if a positive power supply is used with the plus terminal where ground is shown and the ground terminal where the −15 volts is shown. The only disadvantage is that both DVM leads will now be above ground.

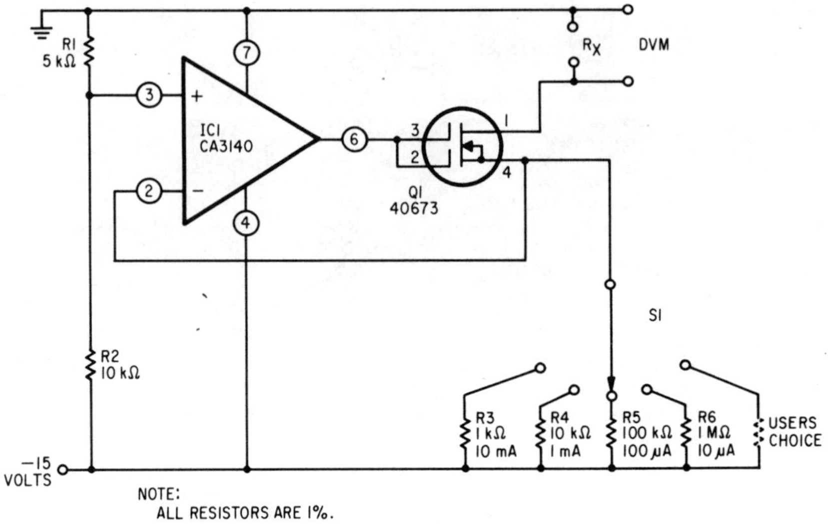

Fig. 1-20 Constant current source schematic diagram.

Parts List

IC1—RCA CA3140 op amp
Q1—RCA 40673 MOSFET (or SK-3065)
R1—5 kilohms, 1%
R2—10 kilohms, 1%
R3—1 kilohm, 1%
R4—10 kilohms, 1%
R5—100 kilohms, 1%
R6—1 megohm, 1%
S1—Single-pole, five-position switch (Centralab PS-101 or equivalent)
Misc.—Printed circuit board, suitable enclosure, terminals, etc.

Construction

The only detail of construction to be mentioned is that the accuracy of the current selected is a function of the accuracies of resistors chosen. One percent resistors are recommended. The printed circuit board is simple and needs no comment. Component location is shown in Fig. 1-21. The circuit will work as well if it is hand wired.

The constant current source may be used as an ohmmeter add-on or in any circuit where a fixed amount of current is needed.

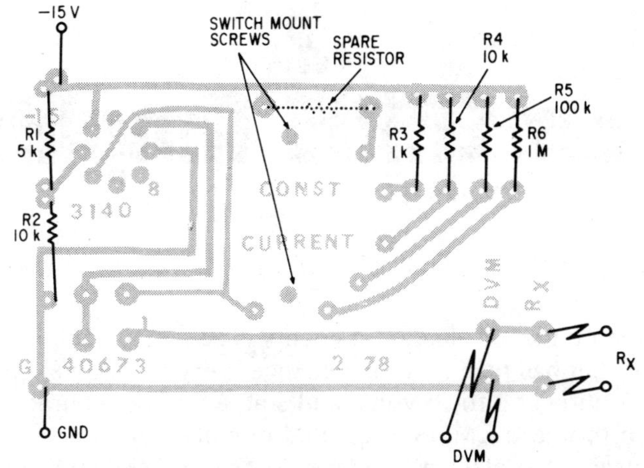

Fig. 1-21 Constant current source component location diagram.

2
Linear CMOS Amplifiers

When in the course of electronic events it becomes necessary to seek a component that has high input impedance, very low power consumption, battery operation at 3 to 15 volts, and stable transfer characteristics, one should turn to linear CMOS integrated circuits.

The uses in the electronic experimenting and amateur radio fields are unlimited. For example, in mobile equipment, where the battery is king, 6 or 12 volt operation is well within the 3 to 15 volts operating range of the CMOS amplifiers.

CMOS is an abbreviation of COS/MOS. The letters come from *CO*mplementary *S*ymmetry/*M*etal-over-*O*xide *S*emiconductor. The idea of a metal-over-oxide semiconductor (MOS) has been around for some time. The MOS is a voltage device and acts much like a vacuum tube but with many advantages including higher input impedance and much lower power consumption.

Figure 2-1 shows an n-channel enhancement MOS. It consists of two n electrodes separated by the p material of the silicon substrate. A metal gate is insulated from the substrate by a thin layer of silicon oxide (glass). It is the very high resistance of this oxide layer that contributes to the very high input impedance at the gate-source terminals.

The space or "channel" between the two n electrodes may be considered to be a variable resistance. With no voltage applied to the gate, the channel resistance is very high and no current flows from source to drain. If the bias voltage on the gate is increased positively toward the drain voltage V_{DD}, it attracts electrons to the surface and lowers the resistance between source and drain electrodes. Current now flows between source and drain. This type of MOS is called *enhancement* because the presence of a proper bias on the gate increases or enhances current flow through the MOS. Input resistance through the gate still remains high because of the insulating oxide between gate and channel. It is on the order of a million megohms.

A p-channel MOS can be formed by replacing the n electrodes with p electrodes, using an n substrate, and reversing the V_{DD} polarity. Now the presence of negative gate bias will have an effect similar to that of positive bias on the n-channel MOS.

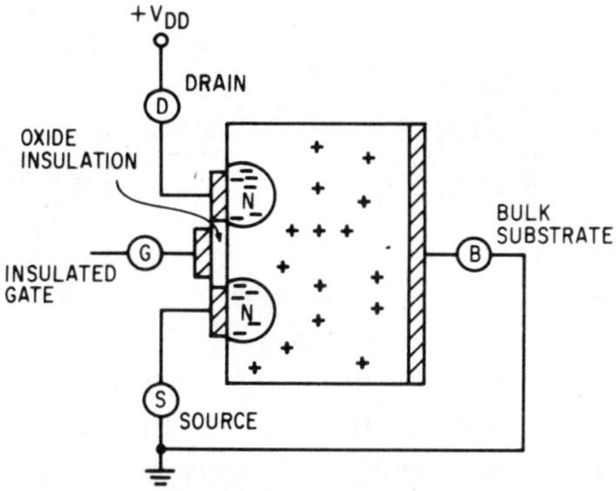

Fig. 2-1 Structure of n-channel insulated gate MOS transistor. The bulk substrate is always connected to the source.

In a CMOS circuit, a p-channel MOS is combined with an n-channel MOS as shown in Fig. 2-2. If the input is held at zero bias (that is, at ground), the gate-to-source voltage of the p-channel device is equal to the supply voltage. It is, therefore, equal to or more negative than the drain of the p device. This causes high current to flow through the p device (low impedance) and results in the output voltage becoming equal to the supply voltage. At the same time, the n-channel device is biased to cut-off forming a high impedance path from the output to ground.

When the input voltage is raised to the supply voltage ($+V_{DD}$) the p channel is turned off and the n channel is turned on. This gives a low impedance path from output to ground and a high impedance path from output to $+V_{DD}$. The output, therefore, approaches zero volts.

By addition of a high resistance (20 megohms) between common gate and common drain connections, the gate bias, with no signal applied, will equal the drain voltage of both MOS transistors and will be at half the

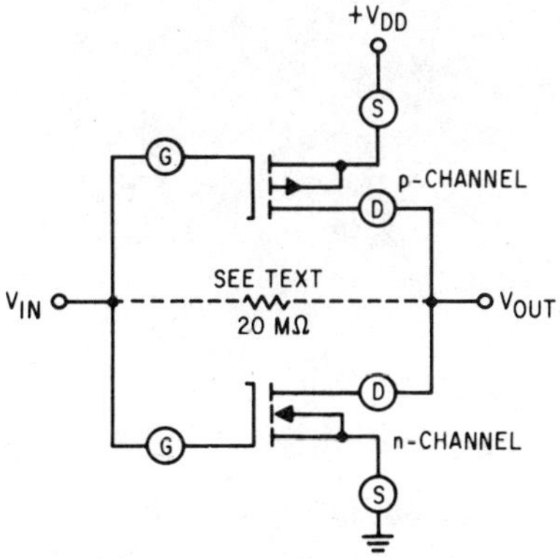

Fig. 2-2 Basic CMOS circuit of the CA3600 transistor array.

$+V_{DD}$ supply voltage. Current will be equally divided between the two MOS transistors. Application of an ac signal to the input will cause an ac signal to appear at the output. Since it takes a small change in input to produce a full swing in output, amplification results. In the CA3600, gains as high as 40 dB can be obtained at 5 volt supply operation.

The first three circuits use the CA3600. This linear device consists of three pairs of CMOS transistors. The independence of each pair permits many uses. A few examples are described in the following three projects. First is a dual-input audio mixer that can be used to feed two microphones into the transmitter for rag-chewing. The second is a two-tone generator for tuning an SSB rig. It is so small, it can be built right into the transmitter or can be hand operated with SSB equipment. And, third, an SSB detector for the all solid-state receiver or as a simple addition to update an older AM receiver is described.

The last four circuits describe active audio filters using the CMOS CA3240 dual operational amplifier. Each circuit is described by itself.

Audio Mixer

The first project, an audio mixer (Fig. 2-3), is the simplest of the three circuits that use the CA3600. With one CMOS IC and seven components, it is possible to combine two separate inputs into a single output.

Fig. 2-3 Audio mixer finished assembly.

Feed this to the transmitter audio input jack and presto, two microphone inputs for rag-chewing when visitors are in the shack. Or use one microphone and your phone patch to clean up the loose wires on the desk.

Circuit Description

Each input is fed to a separate CMOS amplifier. The outputs of these amplifiers are simply tied together and fed into the third amplifier where they mix and appear at its output. The complete schematic appears in Fig. 2-4. What could be simpler?

Construction

As with all the circuits in this book, ease of construction is evident. With the layout shown in Fig. 2-5 the circuit will work as soon as it is finished.

One precaution should be observed with all CMOS devices, digital as well as linear. The input devices are diode protected, but they should still be treated kindly. Be sure power is off before inserting a CMOS device in a circuit. It is also wise to ground yourself when handling the CMOS devices when they are out of the circuit. This is especially true on dry winter days when high electrostatic discharges are common. Care will prevent the need for rebuying blown devices.

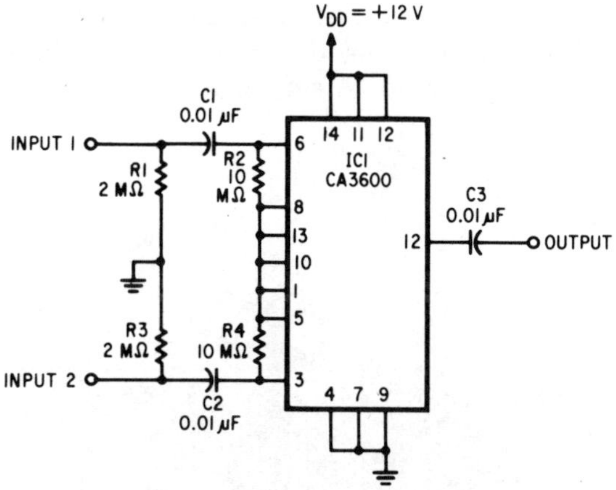

Fig. 2-4 Audio mixer schematic diagram.

Parts List

C1 through C3—0.01-µF, 100-volt ceramic
IC1—RCA CA3600 CMOS transistor array (4007 may be substituted with slight loss of linearity)
R1, R3—2 megohms
R2, R4—10 megohms
Misc.—Printed circuit board, IC socket, terminals, etc.

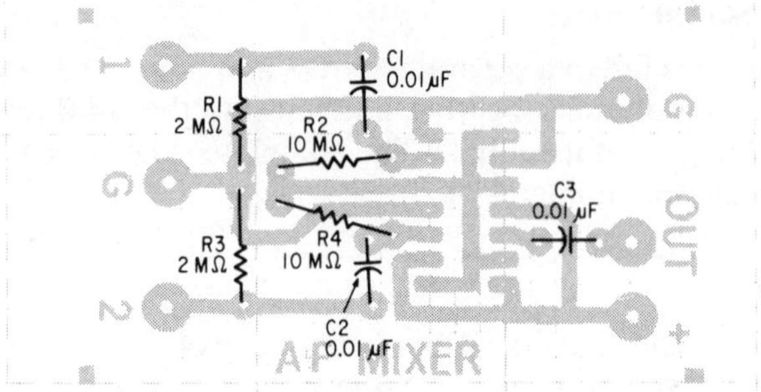

Fig. 2-5 Audio mixer component location diagram.

Two-Tone Audio Oscillator

A slightly more elaborate but very useful tool is the two-tone audio frequency generator for SSB transmitter tuning, shown in Fig. 2-6. The use of two separate RC audio oscillators on 1500 and 1900 hertz mixed in the CMOS pair provides all that is necessary for SSB tuning. Switch S2 allows a choice of one or two tones. The audio frequencies are determined by the RC filters and can be changed by varying the values of R or C. The frequencies shown here should be satisfactory for tuning most any SSB rig.

Fig. 2-6 Two-tone audio oscillator finished assembly.

Circuit Description

The schematic diagram (Fig. 2-7) shows that one CMOS pair is used for each oscillator. The frequency is not critical so 5% resistors and common disc capacitors are in order. The output of each oscillator is mixed in the final CMOS device in the same manner as audio mixing was done in the preceding project.

When used with an oscilloscope, the two tones allow for proper tuning of an SSB transmitter. The method is best described in any recent issue of the *Amateur Radio Handbook* published by the American Radio Relay League, Newington, CT.

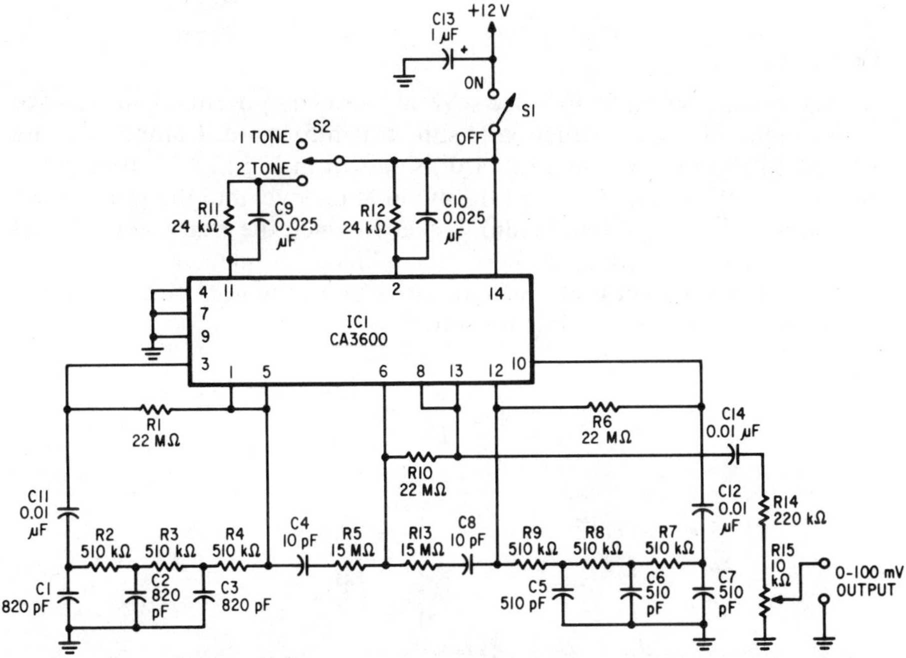

Fig. 2-7 Two-tone audio oscillator schematic diagram.

Parts List

C1 through C3—820-pF, 100-volt ceramic
C4, C8—10-pF, 100-volt ceramic
C5 through C7—510-pF, 100-volt ceramic
C9, C10—0.025-μF, 100-volt ceramic
C11, C12, C14—0.01-μF, 100-volt ceramic
C13—1 μF (The figure shows a monolithic capacitor but an electrolytic capacitor may be used, such as a Cornell-Dubilier PC1-50.)

(*Continued on next page*)

IC1—RCA CA3600 CMOS transistor array (4007 may be substituted with slight loss of linearity)
R1, R6, R10—22 megohms
R2, R3, R4, R7, R8, R9—510 kilohms
R5, R13—15 megohms
R11, R12—24 kilohms
R14—220 kilohms
R15—10 kilohm trim-pot
S1, S2—SPST switch
Misc.—Printed circuit board, IC socket, etc.

Construction

Although the oscillators use several more components than are used in the audio mixer, construction is still straightforward. Components are located on the printed circuit board as shown in Fig. 2-8. A trim pot is provided on the output to adjust the modulation level into the transmitter. Remember, only very little audio is needed since the transmitter has its own amplifier. An output of about 100 millivolts is adequate.

After this project is completed, move on to the next circuit, a product detector to allow for SSB reception.

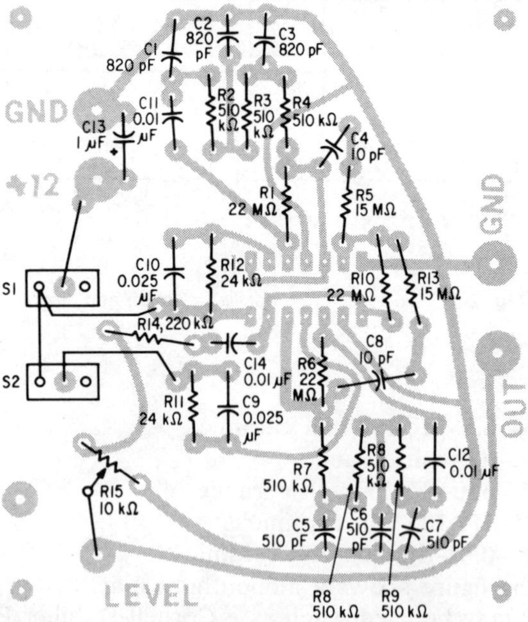

Fig. 2-8 Two-tone audio oscillator component location diagram.

SSB Detector

Product detectors have become standard on all new receivers. This project (Fig. 2-9) can be used in any new solid-state home-brew rig or to add life to that old ''antique'' AM receiver. Or, a used AM receiver can be bought at a low price and then be updated to receive single sideband.

Circuit Description

A crystal oscillator mixes with the receiver IF (intermediate frequency) output signal to give good clean audio on voice or CW. The crystal frequencies must be chosen to match the receiver IF selectivity curve so as to provide an audio bandwidth of about 1750 (350 to 2100)

Fig. 2-9 SSB detector finished assembly.

hertz. For example, a theoretical IF pass-band centered on 456 hertz is shown in Fig. 2-10. The upper-sideband crystal is chosen to be 350 hertz below the 6 dB lower frequency of the IF pass-band. This provides 40 dB or more suppression of the oscillator to prevent it appearing in the audio. The audio pass-band will now be 350 to 2100 hertz. By placing another crystal on the opposite side of the pass-band in the same manner, lower-sideband reception is achieved. Switch S1 in the schematic (Fig. 2-10) provides for upper- or lower-sideband reception.

This circuit has similarities to the preceding two. A crystal oscillator is mixed with the receiver IF frequency to provide detection of the SSB signal. The third CMOS pair is the buffer to the output.

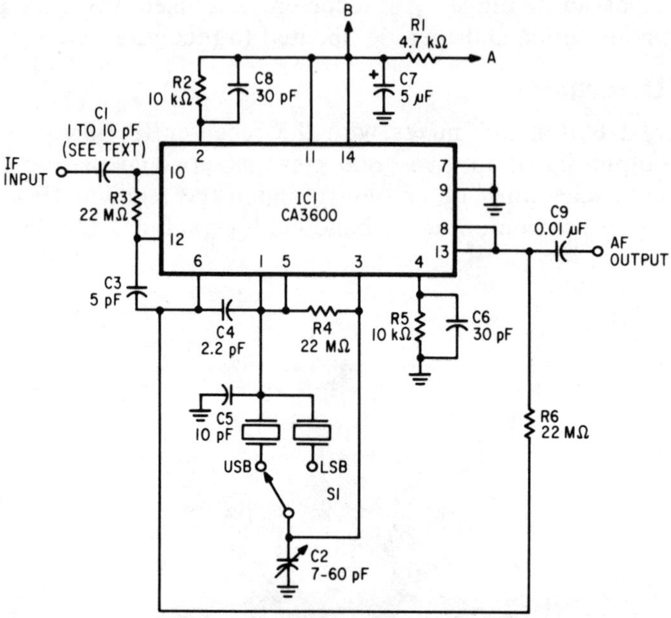

Fig. 2-10 SSB detector schematic diagram.

Parts List

C1—1 to 10pF, see text.
C2—7-60-pF trimmer (ARCO-404 or equivalent)
C3—5pF, 100 volts
C4—2.2 pF (Erie TCZ-2R2 or equivalent)
C5—10-pF, 100-volt ceramic
C6, C8—30-pF, 100-volt ceramic
C7—5-μF, 25-volt electrolytic
C9—0.01μF, ceramic
IC1—RCA CA3600 CMOS transistor array (4007 may be substituted with slight loss of linearity)
R1—4.7 kilohms
R2, R5—10 kilohms
R3, R4, R6—22 megohms
S1—SPDT switch
Misc.—Printed circuit board, terminals, crystal sockets, crystals, etc.

Construction

Make a printed circuit board or wire it by hand using a layout similar to the one shown in Fig. 2-11. The crystal frequencies must be chosen to fall on the outer edges of the receiver pass-band. Allow about 1.5 to 2 kHz either side of the standard IF frequency for the crystal. For example, for a 456 kHz IF, use 454.5 kHz and 457.5 kHz crystals. Many Ham junk boxes will contain surplus crystals that fall in this range. They may also be ordered from any one of several Ham crystal manufacturers. By this method, crystals can be chosen for any receiver IF.

The IF input capacitor C1 should be adjusted to provide a maximum of 0.1 to 0.2 volt peak-to-peak at pin 10 of the CA3600. C2 can be used to trim the oscillator for maximum output. Below about 2 MHz, the product detector will work best with a power supply of 6 volts at point B. If only a 12-volt supply is available, dropping resistor R1 can be used to drop the extra 6 volts by applying the 12 volts at point A. Above 2 MHz, the fall-off of gain with frequency can be overcome by applying 12 volts directly to point B. In this manner, operation up to 4.5 MHz is possible.

Thus we see that the CA3600 is well suited for use in many common Ham circuits where its high input impedance, low power consumption, and three-in-one design allow for compact, useful circuits.

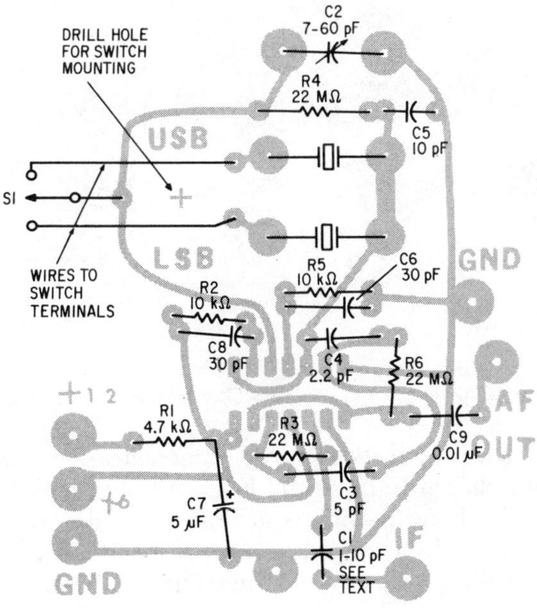

Fig. 2-11 SSB detector component location diagram.

High- and Low-Pass Filters

This and the following project describe four different active audio filters. Each may be used by itself or in any combination to obtain desired results. The filters are called *active* because they use operational amplifiers to produce some small gain as well as to help develop the required pass-band. These filters use only resistors and capacitors to set the frequency response. The operational amplifier chosen is the CA3240 dual CMOS device. Because of the very high CMOS input impedance, large resistor values may be used with a resultant decrease in the physical size of the capacitors. As a result, many capacitors may be located in a small area. This allows several frequencies to be chosen by a switch. First to be described are high-pass and low-pass filters.

Circuit Description

The schematics of both a high-pass and a low-pass filter (Fig. 2-12) are shown in Figs. 2-13 and 2-14. Together they form a band-pass filter. Five cut-off frequencies have been chosen for each filter. The four-pole,

Fig. 2-12 High- and low-pass filter finished assembly.

five-position switch S1 selects the proper capacitors for each range. The frequencies were chosen to be useful in Ham radio, high-fidelity, or any other applications where filtering is desirable. As described later, these frequencies can be easily changed to meet a specific need.

A high-pass filter will pass all frequencies above a chosen cutoff frequency. This is the frequency where the output falls 3 dB. All frequen-

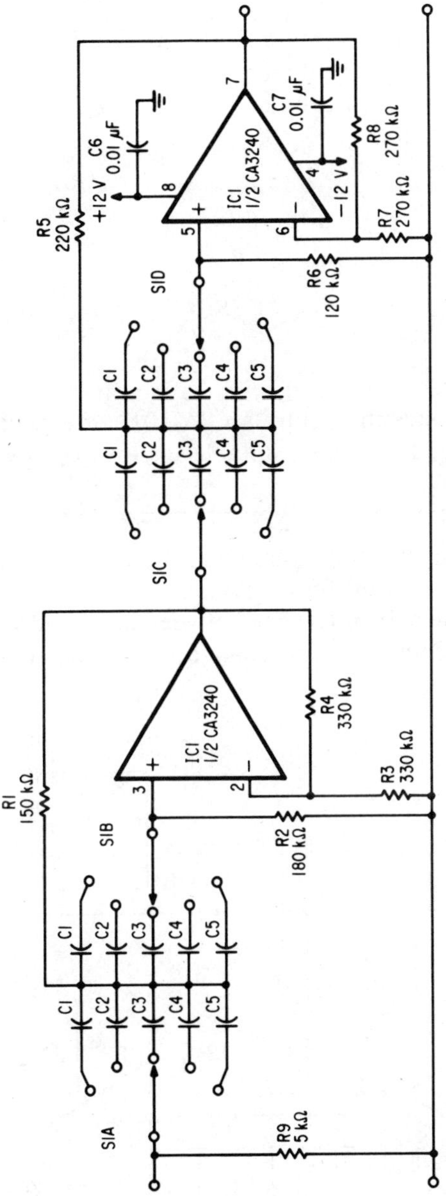

CAPACITOR VALUES	
C1	1000 Hz — 1000 pF
C2	500 Hz — 2000 pF
C3	100 Hz — 0.01 μF
C4	30 Hz — 0.033 μF
C5	15 Hz — 0.068 μF
	LOW FREQUENCY FALL OFF

Fig. 2-13 High-pass filter schematic diagram.

Parts List

C1 (4 req.)—1000-pF mica
C2 (4 req.)—2000-pF mica
C3 (4 req.)—0.01-μF, 80-volt paper tubular
C4 (4 req.)—0.033-μF, 80-volt paper tubular
C5 (4 req.)—0.068-μF, 80-volt paper tubular
C6, C7—0.01-μF, 50-volt ceramic
IC1—RCA CA3240 dual BiMOS op amp
R1—150 kilohms
R2—180 kilohms
R3, R4—330 kilohms
R5—220 kilohms
R6—120 kilohms
R7, R8—270 kilohms
R9—5 kilohms
S1—Four-pole, five-position switch (Centralab PA-1013 or equivalent)
Misc.—Printed circuit board, IC socket, suitable enclosure, power supply, etc.

---

cies above cutoff will pass unchanged. Likewise, the low-pass filter will pass all frequencies below the cutoff frequency.

In the circuits shown, the only difference between high- and low-pass, aside from frequency, is a complete interchange of the capacitors and feedback resistors.

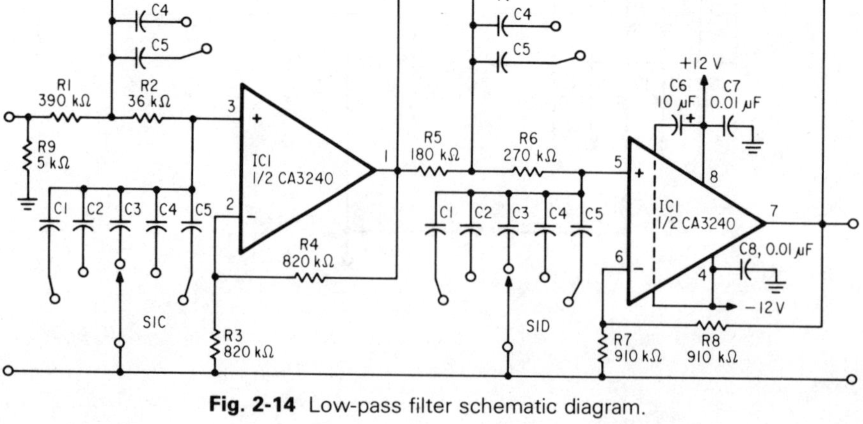

Fig. 2-14 Low-pass filter schematic diagram.

Parts List

C1 (4 req.)—1500-pF mica
C2 (4 req.)—680-pF mica
C3 (4 req.)—330-pF mica
C4 (4 req.)—150-pF mica
C5 (4 req.)—100-pF mica
C6—10-μF, 25-volt electrolytic
C7, C8—0.01-μF ceramic
IC1—RCA CA3240 dual BiMOS op amp
R1—390 kilohms
R2—36 kilohms
R3, R4—820 kilohms
R5—180 kilohms
R6—270 kilohms
R7, R8—910 kilohms
R9—5 kilohms
S1—Four-pole, five-position switch (Centralab PA-1013 or equivalent)
Misc.—Printed circuit board, suitable enclosure, power supply, etc.

In either circuit the cutoff frequency is a function of R and C, but it can most easily be changed by selection of the capacitor set. A dual op amp allows a two-stage filter to be built in a small space. This gives a much sharper fall-off than could be achieved by one alone. The RC values were chosen from graphs in *Manual of Active Filter Design* by John L. Hilburn and David E. Johnson (New York: McGraw-Hill Book Co., 1973).

Construction

The printed circuit board layout (Fig. 2-15) contains both high- and low-pass filters. For best filter response, 5% or better resistors and capacitors should be used. For the lower value capacitors, 5% silver mica are ideal. In the case of the high-pass filter, where large capacitors are required, it becomes necessary to use 10% capacitors. Even so, the filter action is still quite satisfactory.

The use of a four-pole, five-position switch was arbitrary. If there is need for only one specific cutoff frequency, the switches may be omitted and the selected capacitors wired directly into the circuit.

If it is desired to change the cutoff frequency, it is only necessary to change the capacitor values. Remember, frequency is a direct function of the capacitor. To double the cutoff frequency, simply select a capacitor half the size. To reduce the frequency by two, double the capacitor. For example, if 1500 pF is used for 1 kHz, then 680 pF is the approximate standard value for 2 kHz. Note that at each frequency the four capacitors are identical. This should help in matching a set of capacitors. By purchas-

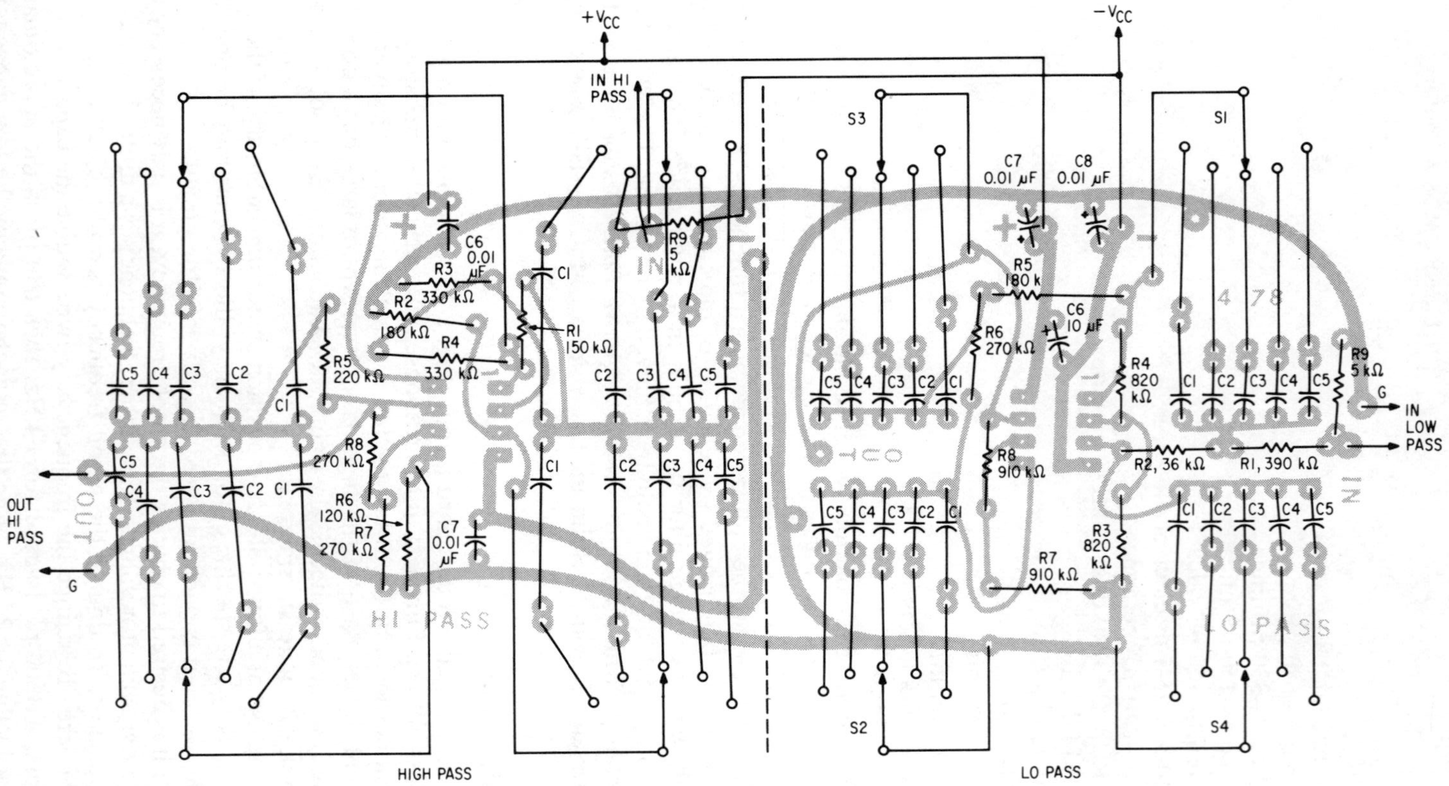

Fig. 2-15 High- and low-pass filter component location diagram.

ing four at once, they most likely will come from one manufacturer's lot. Of course, R and C values can be selected from a larger group by use of a good ohmmeter or bridge for the resistors and a capacitance meter for the capacitors. The exact cutoff frequency is not as important as the rate of fall-off, which is a function of the parts matching.

A band-pass filter is easily obtained by connecting the low-pass output to the high-pass input, thereby putting the filters in series. An inside view of the finished project is shown in Fig. 2-16.

Fig. 2-16 Inside view of the high- and low-pass filter with dual tracking supply in rear.

The next project is a very sharp band-pass filter where only one frequency is desired, such as occurs in receiving Morse code off the air.

Band-Pass and Notch Filters

Described here are two specific sharp filters. The first is a very selective band-pass filter (Fig. 2-17) while the second is the converse; that is, a sharp notch filter (Fig. 2-18). Both have their distinctive use.

In receiving Morse code transmission, as in Ham radio operation, the message is derived from a single tone. With the multitude of stations

Fig. 2-17 Band-pass filter finished assembly.

on the air, all other tones and noises are interference and make copying more difficult.

A sharp selective filter can be made using one CA3240 dual CMOS operational amplifier. By choice of a comfortable pass frequency of about 1000 hertz, Morse reception is greatly enhanced. It might also be noted

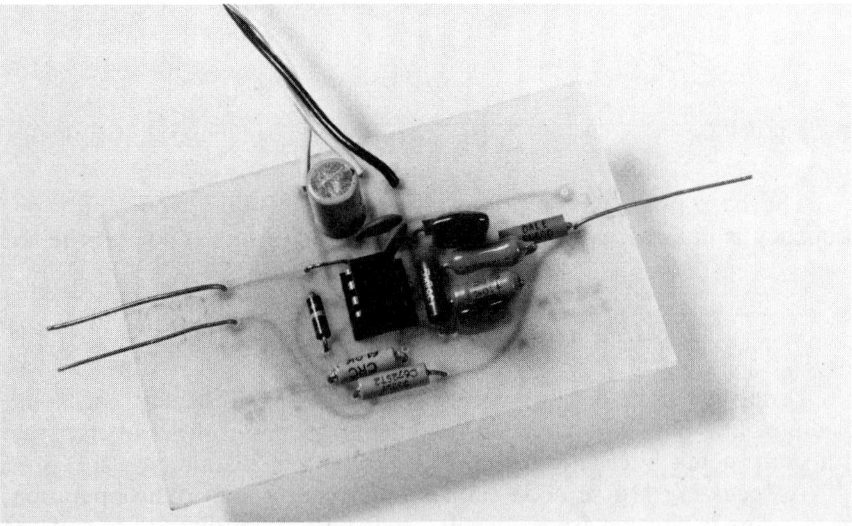

Fig. 2-18 Notch filter finished assembly.

that two such filters set at 2125 and 2295 hertz are ideal for improving 170 hertz frequency shift radioteletype reception.

A notch filter is most useful in removing 60 hertz line noise from audio circuits. Such a filter will pass all frequencies except 60 hertz, where attenuation takes place.

Circuit Description

The schematics for these projects appear in Figs. 2-19 and 2-20. Here again a dual CMOS op amp allows use of relatively small size capacitors, especially in the notch filter at 60 hertz. The filter frequency is a function of the resistors and capacitors used. Again, the frequency can be changed by choice of the capacitor pair. The larger the capacity, the lower the frequency and vice versa.

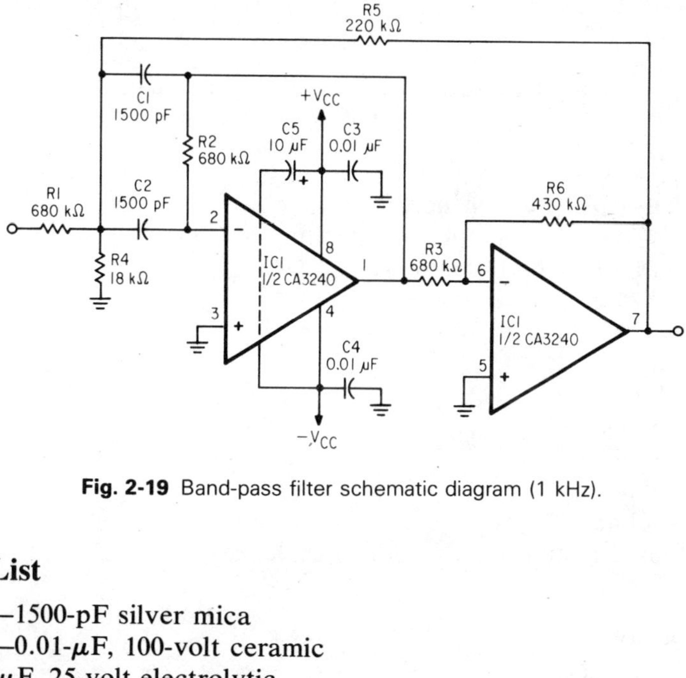

Fig. 2-19 Band-pass filter schematic diagram (1 kHz).

Parts List

C1, C2—1500-pF silver mica
C3, C4—0.01-μF, 100-volt ceramic
C5—10-μF, 25-volt electrolytic
IC1—RCA CA3240 dual BiMOS op amp
R1 through R3—680 kilohms
R4—18 kilohms
R5—220 kilohms
R6—430 kilohms
Misc.—Printed circuit board, IC socket, suitable enclosure, terminals, etc.

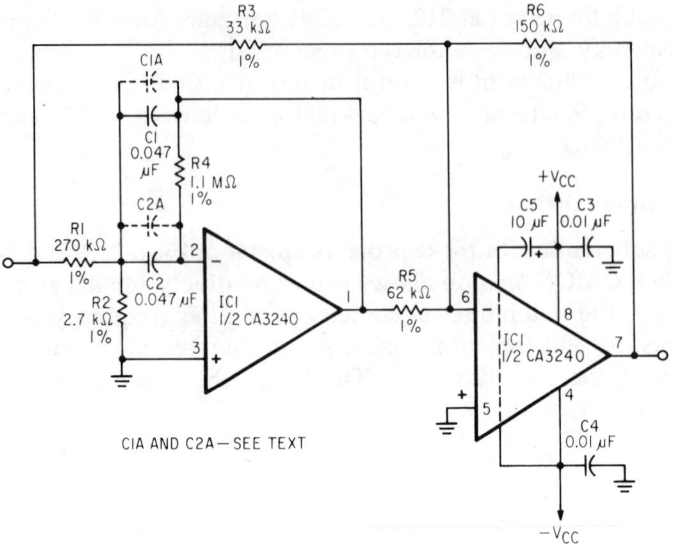

Fig. 2-20 Notch filter schematic diagram (60 Hz).

Parts List

C1, C2—0.047-μF, 100-volt polyester
C1A, C2A—Suitable adjustment capacitors for best notch; see text
C3, C4—0.01-μF, 100-volt ceramic
C5—10-μF, 25-volt electrolytic
IC1—RCA CA3240 dual BiMOS op amp
R1—270 kilohms, 1%
R2—2.7 kilohms, 1%
R3—33 kilohms, 1%
R4—1.1 megohms, 1%
R5—62 kilohms, 1%
R6—150 kilohms, 1%
Misc.—Printed circuit board, suitable enclosure, terminals, etc.

Construction

The photographs (Figs. 2-17 and 2-18) show the simplicity of the printed circuit boards needed for these circuits. Component locations are shown in Figs. 2-21 and 2-22. For the selective filter, the values shown will give a sharp pass-band centered around 1000 hertz, which is a comfortable tone for Morse reception. As before, this frequency can be changed by choice of different capacitor values. The exact frequency is not as important as the matching of the RC values. Therefore, 5% components should be used or, as described in the previous project, values can be selected.

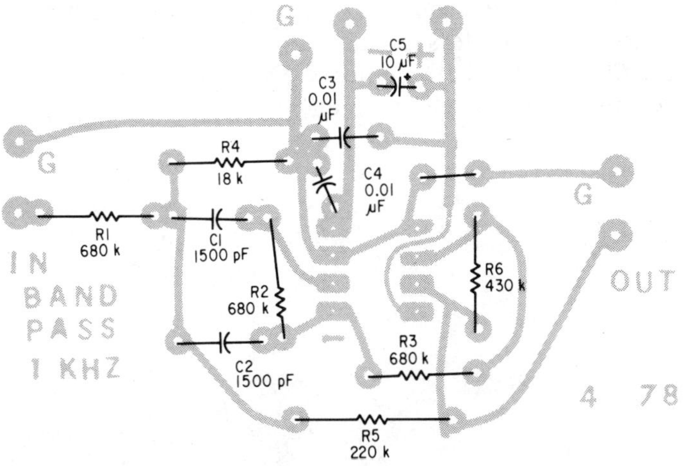

Fig. 2-21 Band-pass filter component location diagram.

The notch filter at 60 hertz should be tuned as close to this frequency as possible to provide maximum attenuation of line hum. Therefore, extra contacts are provided on the printed circuit board to allow use of spare capacitors in parallel with the main ones. Best attenuation at 60 hertz is achieved by using a 60-hertz signal of 1 volt or so and hand selecting the "tune-up" capacitors. Such a signal can be obtained from a signal generator or from a low voltage filament transformer. An oscilloscope attached to the output is most useful in watching the output attenuation while tuning the filter. An ac voltmeter may also be used.

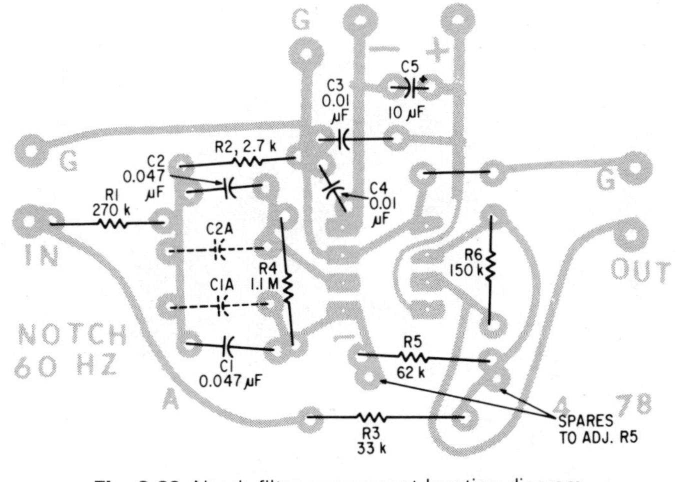

Fig. 2-22 Notch filter component location diagram.

The added capacity should be equally divided between the two main capacitors to keep the filters balanced.

These filters will greatly improve Morse code reception and eliminate 60-hertz hum in any audio equipment.

3
Passive Circuits

After all the labor of love that went into building any or all of the earlier active circuits, it will be nice to build some passive devices that are equally useful.

This section covers a resistance substitution box, a decade box, a Wheatstone bridge, and an attenuator. Each has its place on the workbench or in operating equipment. Don't let their simplicity detract from their usefulness.

Resistance Boxes

A resistor substitution box and a decade box are the next projects to be described. While neither uses any active transistors or integrated circuits, they are most useful in designing circuits that do.

Often, even after all the design calculations are made and the circuit wired, operation is not as good as was expected. Changing resistor values can often tremendously improve operation. Even a circuit that works fairly well can be optimized by small variations in component values. One way to do this is to solder, unsolder, and resolder resistors until the best value is found. A much easier method is to clip a resistor substitution box (Fig. 3-1) into the circuit and quickly turn the switches. No solder, less mess, and no damage to the printed circuit board.

Another useful tool is one where the resistors are chosen so that each switch provides a decade of resistance in 10 steps. Such a device is shown in Fig. 3-2. With this equipment, really fine tuning is possible. The smallest step can be one ohm in a total of 100 thousand. This box is also useful in a resistor bridge for measuring other resistor values to very close tolerance. Its use with a Wheatstone bridge will be described in the latter project.

Circuit Description

As can be seen (Fig. 3-3) the resistor box circuit is simple and a detailed description is not needed. Each switch chooses one of 12 resistors. A separate set of binding posts for each switch allows several resistors to be used at once. By connecting various resistors in series (or

Fig. 3-1 Resistor substitution box finished assembly.

parallel), fine adjustment is possible. The resistors in each section are the standard 10%, 1 watt values that are readily available.

The decade box circuit (Fig. 3-4) is a little more complicated because it was desired to save on the cost of the 1% resistors used. Only four resistors are used per decade. This allows a considerable savings on the cost of resistors. By proper switch wiring, the resistors are put in series to give a unit step as the switch is rotated. For example, to obtain 6 ohms, the 4- and 2-ohm resistors are in series, while for 7 ohms the 4- and 3-ohm resistors are used.

Fig. 3-2 Resistance decade box finished assembly.

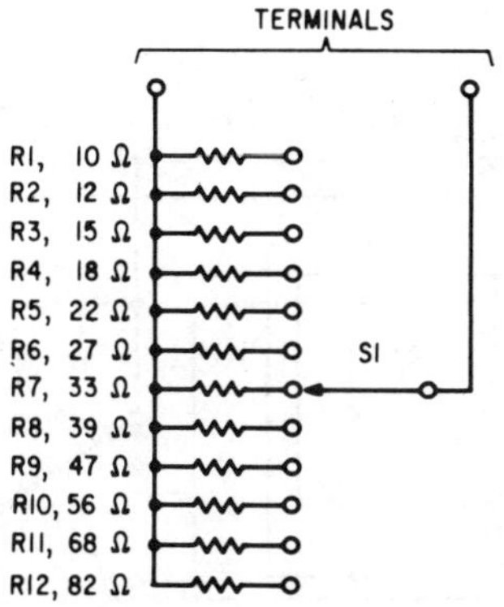

Fig. 3-3 Resistor substitution box schematic diagram showing one of six sections. For the remaining sections, multiply the values shown by factors of 10.

Parts List

R1—10 ohms
R2—12 ohms
R3—15 ohms
R4—18 ohms
R5—22 ohms
R6—27 ohms
R7—33 ohms
R8—39 ohms
R9—47 ohms
R10—56 ohms
R11—68 ohms
R12—82 ohms
(For each additional range, multiply the values given above by factors of 10.)
S1—Single-pole, 12-position switch (1 req. for each range)
Misc.—Suitable enclosure, terminals, etc.

Construction

The photographs show the simple, straightforward construction. For the substitution box (Fig. 3-5), the barrel construction gives a sturdy short-lead structure that will take much abuse. The panel is labeled with Datak rub-on digits, but it could be hand lettered with pen and black ink. The aluminum panel is first rubbed with steel wool to remove oil and fingerprints. After lettering is complete and checked, a light coating of clear spray will help preserve the appearance. The decade box wiring is a bit more complex because of the switch circuit used. Each decade is wired the same except for the values of resistors used. Also, each decade is series wired to the next so only one set of output posts is required.

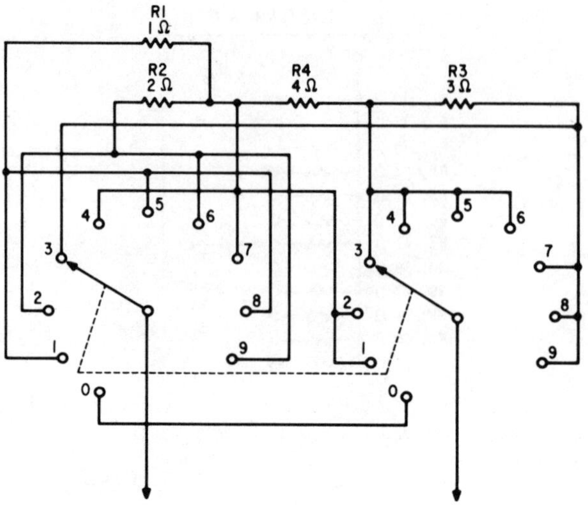

Fig. 3-4 Resistance decade box schematic diagram showing one of six sections. For each range, multiply the four resistance values by factors of 10.

Parts List

R1—1 ohm, 1%
R2—2 ohms, 1%
R3—3 ohms, 1%
R4—4 ohms, 1%
(For each additional range, multiply the values given above by factors of 10).
S1—Double-pole, 10-position switch (1 req. for each range)
Misc.—Suitable enclosure, terminals, etc.

Maximum current for ½-watt resistors

Range	*Current, mA*
1 through 9	200
10 through 90	100
100 through 900	35
1000 through 9000	11
10,000 through 90,000	3.5

The only precaution to be taken in using either box is the choice of resistor wattage. The 1-watt resistors used in the substitution box are heavy duty and suitable for use in most solid-state circuits except those incorporating power devices.

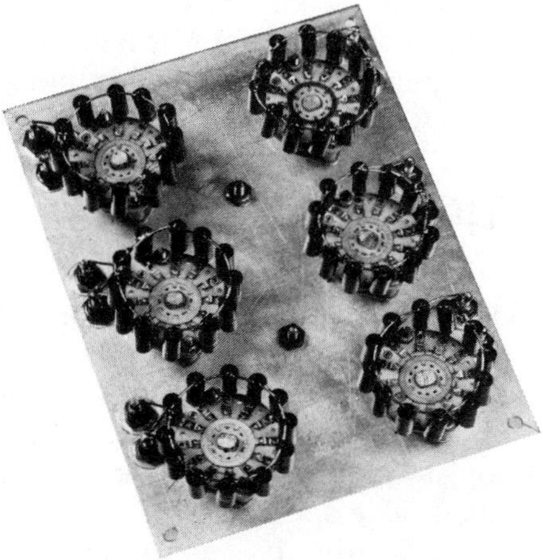

Fig. 3-5 Inside view of the resistor substitution box.

The decade box (Fig. 3-6) requires a bit more care. The maximum allowable current for each decade is shown in the parts table for ½-watt resistors because these are most easily obtained. If higher wattage 1% resistors can be obtained, then higher currents can be used. The box shown contains 2-watt, 1% resistors because these happened to be available at construction time. It also covers six decades from 0.1 ohm in 0.1

Fig. 3-6 Inside view of the resistance decade box.

ohm steps to 100 kilohms in 10 kilohm steps. Choose whatever set of decades are most useful to you. The four decades from the 10 ohms per step through 10 kilohms per step will fill almost all test needs.

Wheatstone Bridge

If the title sounds familiar, it should. Sir Charles Wheatstone developed this circuit about 125 years ago. Don't let its age fool you. The simple bridge to be described (shown in Fig. 3-7) can be used with the decade resistance box to measure resistors to greater accuracy than can be done with the hobbyist voltmeter or VTVM. The measurement accuracy is a function of the bridge accuracy.

The decade box project uses 1% resistors so an accuracy of about this amount can be obtained. The average ohmmeter is only good for 5% accuracy. The bridge also provides an excellent means of matching resistors. One resistor is attached at the decade box terminals. Other resistors are then tried at the unknown resistor terminal. The closer the meter is to null, the closer the resistors match.

Fig. 3-7 Wheatstone bridge finished assembly.

Circuit Description

An examination of the schematic (Fig. 3-8) will show that there are two voltage dividers across the power supply. The top leg of each is 1 kilohm while the bottom legs are the decade box and unknown, respectively.

The meter measures the difference between the two dividers. When the two resistors are equal, the voltage across the zero centered meter will be zero and the meter will be nulled. A meter shunt resistor is left in the circuit until the meter reads close to center. It is then removed and a final fine adjustment is made.

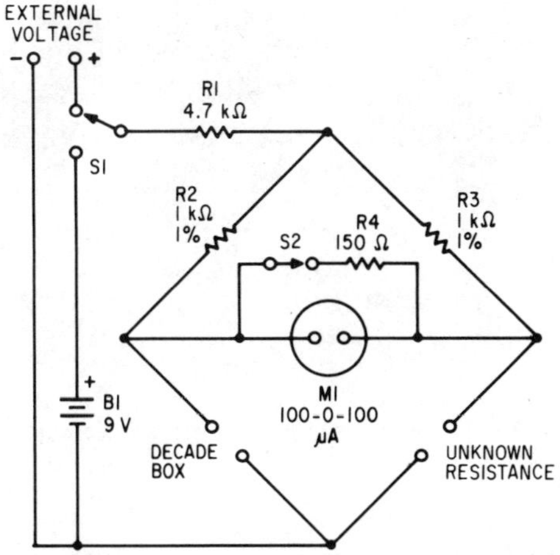

Fig. 3-8 Wheatstone bridge schematic diagram.

Parts List

B1—9-volt battery
M1—100-0-100-μA microammeter
R1—4.7 kilohms
R2, R3—1 kilohm
R4—150 ohms
S1—SPDT switch
S2—SPST switch, normally closed
Misc.—Suitable enclosure, terminals, etc.

Construction

Construction of the Wheatstone bridge is straightforward and needs little comment. All components except the battery are mounted on the front panel of the enclosure, as shown in Fig. 3-9. The meter shunt switch S2 should be normally closed with a momentary open position. This will prevent damage to the meter when testing is started on an unknown resistor. The meter may be any zero center model, with 100 μA each side of center a good choice. The more sensitive the movement, the finer the null point and the more carefully it must be handled. Do not push the shunt switch till the meter is almost fully nulled. Switch S2 allows use of the fixed 9-volt transistor battery or an external supply if desired. When not in

Fig. 3-9 Inside view of the Wheatstone bridge.

use, the battery is protected as long as no resistors are attached to the terminals. As an added precaution, set S1 to EXTERNAL to disconnect the battery.

Remember, this bridge can be used to match resistors for use in such circuits as the audio filters described previously.

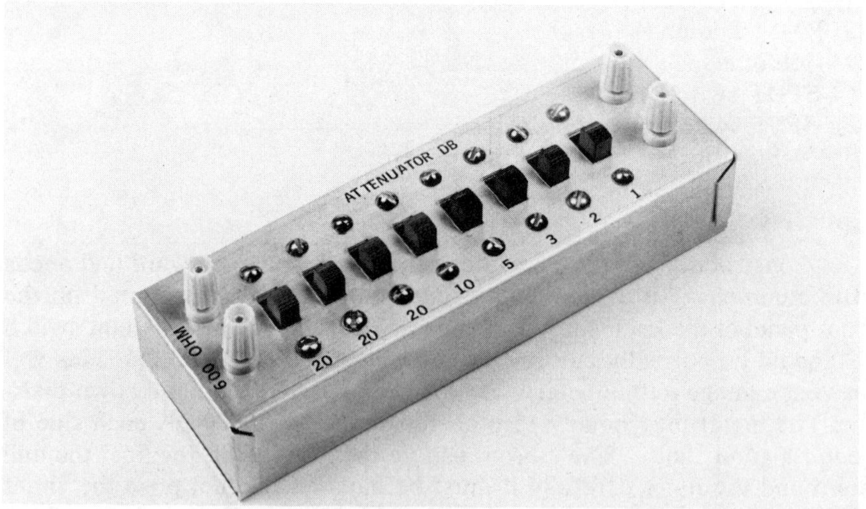

Fig. 3-10 Passive attenuator finished assembly.

Passive Attenuator

What will I ever do with an attenuator? Why, you'll do with the attenuator exactly what its name says. Sometimes we don't realize how much can be done with such simple devices. Any audio volume control is a continuous attenuator. So is a simple voltage divider. A switching attenuator has one advantage: It reduces a signal by a measured amount. The attenuator described is shown on the previous page in Fig. 3-10.

If it is calibrated in dB, it may be used with an oscilloscope to compare the amplitudes of two signals on the screen with a direct answer in decibels. Simply pass the larger signal through the attenuator, throw switches until the two signals are equal, and read the answer on the attenuator.

It may also be used to adjust the input to stereo amplifiers. One 20-dB switch will double or half the signal.

Parts List

R1, R2, R4, R5, R7, R8—750 ohms, 1%
R3, R6, R9—3010 ohms, 1%
R10, R11—1210 ohms, 1%
R12—825 ohms, 1%
R13, R14—2210 ohms, 1%
R15—357 ohms, 1%
R16, R17—3570 ohms, 1%
R18—221 ohms, 1%
R19, R20—5110 ohms, 1%
R21—130 ohms, 1%
R22, R23—5620 ohms, 1%
R24—61.9 ohms, 1%
R25—600-ohm external resistor
S1 through S8—DPDT switch
Misc.—Suitable enclosure, terminals, etc.

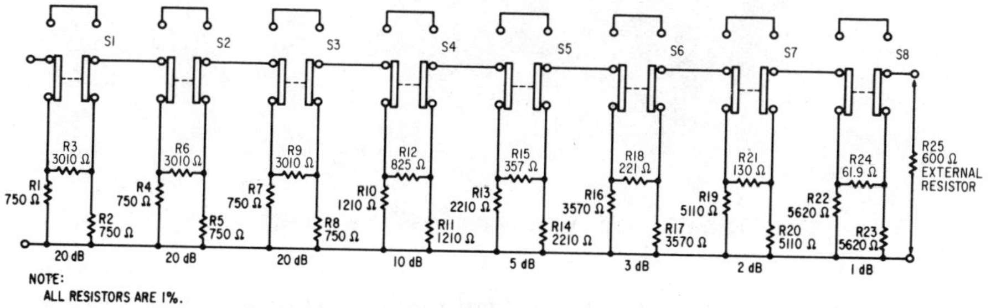

Fig. 3-11 Passive attenuator schematic diagram.

Circuit Description

Each stage of the attenuator is made of three resistors. In effect, two resistors shunt the third. The values of the shunt pair determine its tap and the amount of attenuation. Since these values can be accurately calculated, the amount of attenuation can be chosen and the switches properly marked.

The box shown has eight sections (see Fig. 3-11 on the previous page). Three are designed for 20 dB, one for 10 dB, and one each for 5, 3, 2, and 1 dB. By choice of switches, any value from 1 to 81 dB may be selected. This attenuator is designed for a 600-ohm audio line and therefore should have a 600-ohm load or shunt on the output to read correctly.

Construction

For best accuracy, 1% resistors should be used. However, 5% values are still reasonably good. Mount the eight switches and input and output terminals. Run a bare wire from one input terminal to one output terminal. Use this wire as the common tie line for two of each three-resistor set. The third resistor is wired across the switch arms. In the OUT position, the switches bypass the attenuator networks. In the other position, the network is used to attenuate. Figure 3-12 shows an inside view of the attenuator enclosure and means of mounting the resistors.

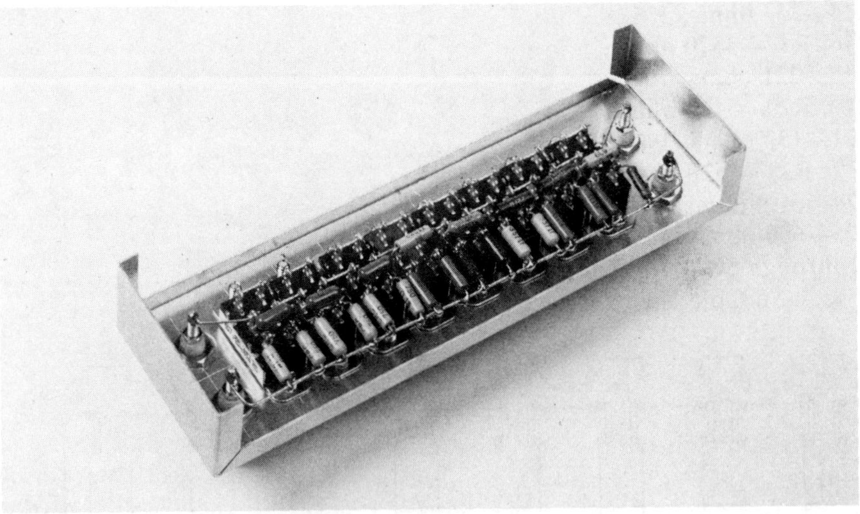

Fig. 3-12 Inside view of the passive attenuator.

With this tool, the workbench will really start to look like a professional lab.

4
Instruments

The hobbyist needs two sets of construction tools. The basic mechanical tools come first. Without the pliers, screwdriver, soldering iron, and so on, no electronic equipment could be built. Without the second set, the electronic tools themselves, no troubleshooting and adjustment could be made.

To some, this part of the book will be the most useful. Starting with a low cost basic dc DVM using the latest ICs, several add-on items are described. These allow the DVM to be used as an ac-voltage instrument and to have semiautomatic ranging. When used with the constant current source described in Chap. 1, ohmmeter operation is possible.

Also described is a low cost frequency counter together with a prescaler to extend its range. A novel RF detector is discussed, as is a digital probe.

These devices are most useful, in fact almost essential, for troubleshooting new projects or servicing older equipment.

Multipurpose RF Detector

A sensitive RF detector is a very useful device in the home as well as on the electronic workbench. Suppose the family is watching TV and suddenly there is RFI (radio-frequency interference) and a distorted picture. Was the cause the furnace motor, the wife's hair drier, or some other of the dozens of electrical appliances in the house? Maybe the RFI is from a neighbor's house. Use the RF detector (shown in Fig. 4-1) to find the source.

Perhaps a device has just been built and it won't work. Without an oscilloscope it is impossible to tell if the oscillator stage is operating. Maybe a power supply that has just been completed is causing strange sounds in your hi-fi system. Could it be oscillating at a low frequency? Maybe the new Ham antenna needs tuning for maximum output. Here, this RF detector can serve as a broad-band field strength meter. It was built to trace down oscillations, both those that are wanted and those that are not wanted. Its usefulness is limited only by the imagination of its operator.

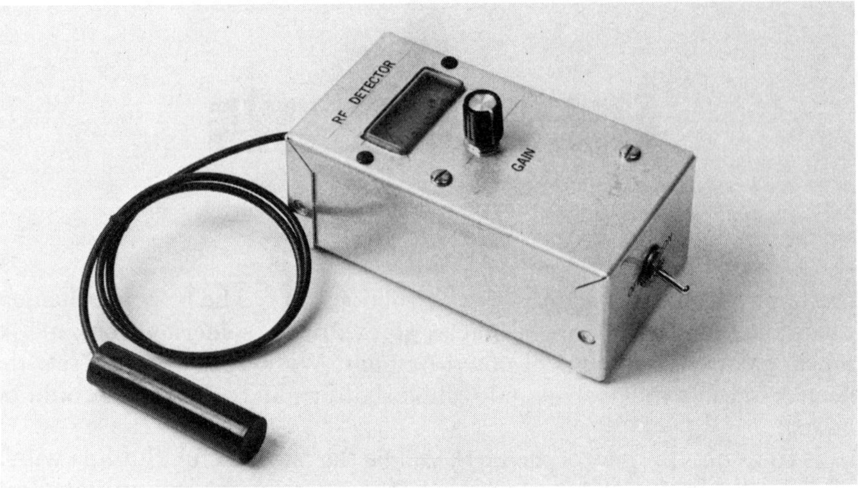

Fig. 4-1 RF detector finished assembly.

Circuit Description

Only a few microwatts of unwanted oscillation can cause severe RFI. Desired oscillations of similar level are often needed to make a circuit do its job. A simple diode detector has poor sensitivity: It takes watts of power output to obtain an indication.

The RF detector circuit (Fig. 4-2) overcomes this problem by use of a high gain, micropower operational amplifier. The CA3078 requires about 15 microamperes of standby current to be operational. When RF is detected, the increase in battery current is only the few more microamperes necessary to move the meter needle. As a result, a battery will last almost its full shelf life when used in this device.

The RF probe consists of a special pickup coil and diode detector. The output is fed to the CA3078 op amp where it is amplified as much as several thousand times to provide an operating voltage for the meter. The amplifier is connected in the noninverting mode with its gain adjusted by the 1 megohm pot on the front panel. Because of the low power requirement, only two size AA dry cells are needed for power. In order to null the op amp input offset, two 1N914 diodes and a 200 kohm resistor are series connected across the 3-volt battery supply.

The forward voltage drop of the diodes D1 and D2 provides a reference voltage point of 1.4 volts or approximately one-half of the 3-volt supply. The inverting input of the op amp is returned to this point. This allows use of a trim pot and series resistor connected to the noninverting, signal input of the op amp to null any offset voltage. A Zener diode, D3, across the input is used to prevent overdrive to the operational amplifier.

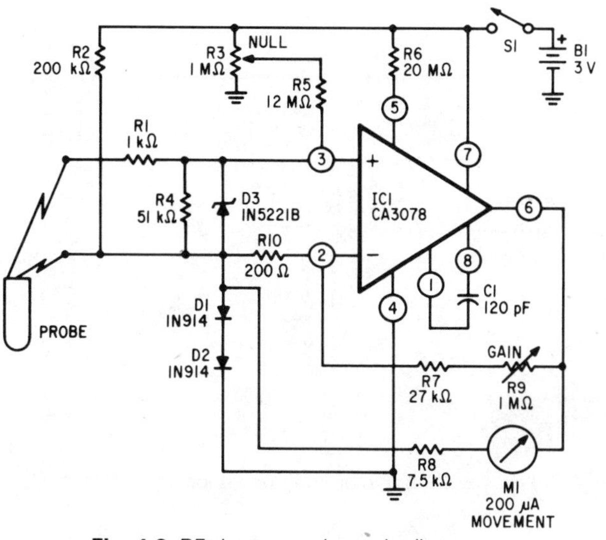

Fig. 4-2 RF detector schematic diagram.

Parts List

B1—AA-size dry cells, 2 req.
C1—120-pF, 100-volt ceramic
D1, D2—1N914 silicon diode
D3—1N5221B Zener diode
IC1—RCA CA3078 micropower op amp (or SK-3566)
M1—200-μA microammeter, miniature
PROBE—Dycom SK-1 RF probe (contact Dynamic Communications, 948 Avenue E, Riviera Beach, FL 33404
R1—1 kilohm
R2—200 kilohms
R3—1-megohm trim-pot (CTS X201R-105 or equivalent)
R4—51 kilohms
R5—12 megohms
R6—20 megohms
R7—27 kilohms
R8—7.5 kilohms
R9—1-megohm gain potentiometer, miniature
R10—200 ohms
S1—SPST switch
Misc.—Suitable aluminum enclosure, printed circuit board, etc.

Construction

The printed circuit board provides a compact, stable layout of parts (Fig. 4-3). An inside view of the finished project is given in Fig. 4-4. Most

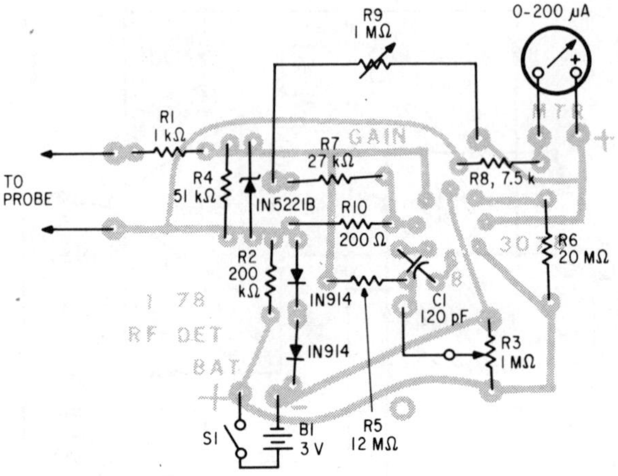

Fig. 4-3 RF detector component location diagram.

of the components are standard, but a comment should be made about the probe and the meter.

The RF probe is manufactured by Dynamic Communications, whose address is given in the Parts List. Its special value lies in its broad-band characteristic. It will sense oscillations from a few hundred kilohertz to several hundred megahertz. Obviously no tuned circuits are needed, and one does not have to know the oscillation frequency to detect it.

Fig. 4-4 Inside view of the RF detector.

A surplus 200-microampere meter was used in this project. Any sensitive meter can be used. The more sensitive the meter, the less the battery drain when detecting a signal. The resistor in series with the meter may have to be changed if a different range meter is used. It should be adjusted to a value that will prevent the meter from slamming the full-scale end on a strong signal.

Because of their long life expectancy, the batteries were soldered into the circuit.

After the detector has been assembled, hold the probe in front of a TV screen. A full-scale swing of the meter needle should be observed even with the gain set low. With full gain, the small power output of an RF signal generator will cause full-scale swing. Remember, this device can also be used as a field strength meter for tuning that Ham transmitter and antenna without the need for a tuned circuit in the detector.

Portable Digital Voltmeter

The digital voltmeter (DVM), incorporating complicated circuits and many components, was once the rich man's tool. Recent developments in integrated circuits have put the complex circuits on a chip and have made it very simple for the hobbyist to construct his own DVM at a reasonable cost. This DVM (Fig. 4-5) has several features besides low cost.

First, there is 5-volt rechargeable battery operation for portability. Second, automatic blanking of the digits after 20 seconds conserves bat-

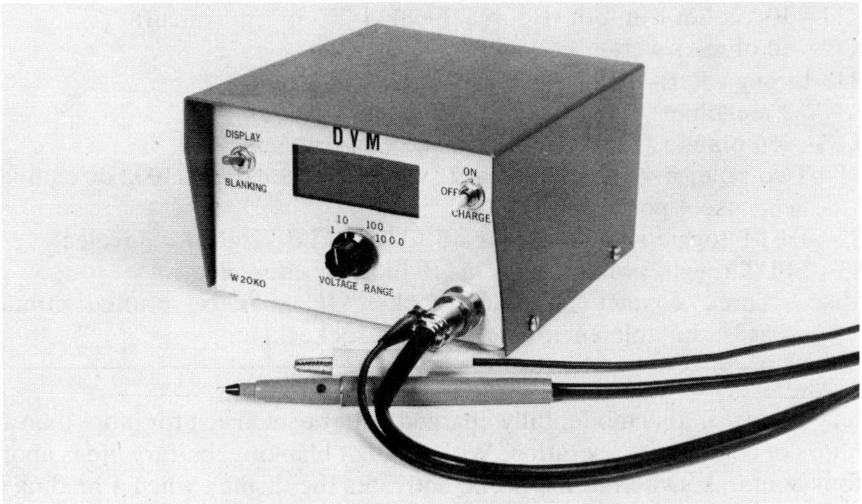

Fig. 4-5 Portable DVM finished assembly.

Parts List

B1—5-volt, NiCd battery pack
C1, C10—0.01 μF, 1 kilovolt (Centralab DD103 or equivalent)
C2—470-pF mica
C3—0.27-μF, 80-volt tubular
C4—4.7-μF nonpolarized (Radio Shack 272-498 or equivalent)
C5 through C9—0.01-μF, 100-volt ceramic
D1—1N270 germanium diode
I1 through I3—FND 507 or FND 510 displays
IC1—RCA CA3130T BiMOS op amp
IC2—RCA 3161 BCD-to-7-segment decoder/driver
IC3—RCA CA3162 analog-to-digital converter
IC4—4047A monostable/astable multivibrator
IC5—4071B quad 2-input OR gate
LED, miniature
Q1 through Q3—2N4036 pnp power transistor
R1—9.09 megohms, 1%
R2—909 kilohms, 1%
R3—90.0 kilohms, 1%
R4—10 kilohms, 1%
R5—22 megohms
R6—2.7 kilohms
R7—9.1 kilohms
R8—2 kilohms
R9—100-kilohm trim-pot (Bourns 3068P-1-203 or equivalent)
R10—50-kilohm trim-pot (Bourns 3068P-1-203 or equivalent)
R11—10-kilohm trim-pot (Bourns 3068P-1-203 or equivalent)
R12—10 ohm, 1 watt
R13 through R16—12 kilohms
R17—2 megohms
R18—180 ohms
S1—Two-pole, five-position rotary switch (Centralab PA-1002 or equivalent, use 4 positions)
S2—DPTT toggle switch, center off (Alco MTB-206P or equivalent)
S3—SPDT toggle switch (Alco MTB-106D or equivalent)
Misc.—Three contact plug and socket, IC sockets, printed circuit boards, suitable enclosure, charger socket, etc.

---

tery power. In this mode, fully charged batteries will last for more than 12 hours of continuous operation. Without digit blanking, battery life is about four hours. A switch in the probe activates the display when a reading is needed. Third, the last-measurement-hold feature allows the hobbyist to probe a reading, remove the probe, and still have the display show the

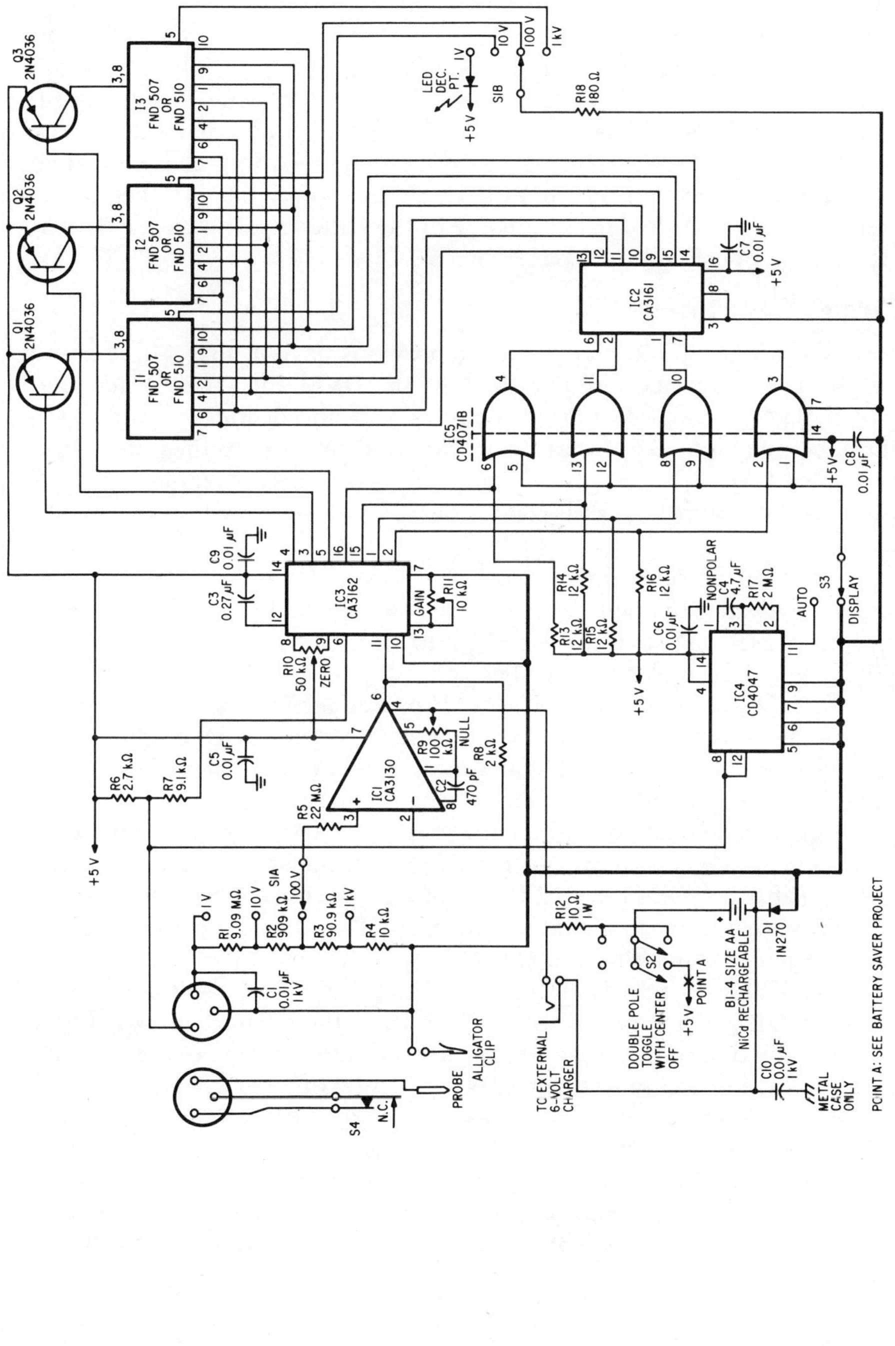

Fig. 4-6 Portable DVM schematic diagram.

voltage read. This is especially useful when measurements are made on printed circuit boards where components are closely spaced. With the hold feature, the eyes can be kept on the probe. Accidental destructive shorts will not be caused because the probe slipped while the user was trying to read the meter. Probe the terminal to be measured, push the button on the probe, lift the probe free, and read the voltage. Fourth, there is automatic polarity. If the probes are reversed, a two-digit readout will appear with the third left-most display showing a minus sign. Fifth, over-range is displayed as three letter Es in positive polarity and as three minus signs in reverse. Sixth, the external plug-in battery charger can be left attached for bench operation from the ac power line. In this mode, there will also be a slow trickle charge of the batteries. The display blanking may be by-passed by a panel switch.

Circuit Description

The DVM circuit, shown on the previous page in Fig. 4-6, is based on two new integrated circuits, the CA3162 DVM and CA3161 decoder/driver, that provide full voltage measurement and decoding to drive the digits. A complete DVM can be constructed using only these two ICs. Two more CMOS ICs add the display hold and blanking features. CMOS devices draw so little current that they have negligible effect on battery drain.

The voltage being measured is applied to a standard attenuator to provide four measurement ranges: 1, 10, 100, and 1000 volts. The CA3130 op amp is used to provide a very high input impedance so as not to load either the circuit under test or the attenuator.

The CA3162 DVM converts the measured analog voltage into a digital equivalent that appears as a binary coded decimal (BCD) output. Since the BCD output must provide different information for each of three digits, a multiplex system is provided. In multiplex readout, the equivalent segments of each digit are tied together. However, the digits are turned on in sequence, one at a time, synchronized with the specific digit information available from the CA3162 BCD output. A conversion is necessary to provide seven-segment digital information. This decoding from BCD to seven segment is done by the CA3161.

The CD4047 and CD4071 are used to provide the display blanking. The four OR gates between the DVM IC and the decoder are used as a switch. Blanking of the digits occurs when the four inputs to the decoder are all at logic 1. All seven segments are turned off. The characteristic of an OR gate is that if any input line goes to logic 1, the output will also be a logic 1. One input line of each of the four OR gates are tied in common to the CD4047 monostable timer output. When the timer is activated by the probe switch S4, its output goes low to logic 0. In this state, the other input line of each OR gate can switch the output high or low depending on the signals available from the DVM device. After about 25 seconds (a time set by RC), the timer output will swing high. Now, with all four gates having one input high, their outputs are also high. The decoder reads this as a logic 1111 BCD input and turns all the segments off. They will remain off until the probe switch is again pressed. This switch momentarily removes a ground from the timer input, resets the timer, and turns the digits on. This feature may be by-passed by use of switch S3. In the display mode, the common OR gate inputs are tied to ground and the display digits stay lit.

The hold feature to permanently display the last reading is also actuated by the probe switch. When the junction of R6 and R7 is allowed to go above ground by opening the probe switch, the 1 volt that appears at pin 6 of the CA3162 causes the hold feature to operate. When the switch is released, the meter returns to the normal state.

Construction

Two printed circuit boards are used to provide a reasonably sized panel display in a small package. The layout allows connections between the boards without the use of a cable. The range, display, and power switches are soldered directly to the vertical board (Fig. 4-7A) and are also used to attach the printed circuit boards to the front panel. Component location on the horizontal board appears in Fig. 4-7B. A red filter and thin lucite panel are held in place by the two toggle switch mountings. The red filter makes for a neat appearance and hides the wiring behind the panel.

Construction of the probe switch may take a bit of imagination on the part of the hobbyist. Figure 4-8 shows one way it can be done. The entire package fits into an empty felt-tip pen case. If the hobbyist does not mind using a larger diameter probe, a tiny switch might be found in the surplus market. The switch arms are made from two leaves of a small relay that was disassembled. The use of nylon screws simplifies insulation between the leaves. A small nylon screw glued to the movable switch arm provides a safe button which passes through the case. First assemble the switch, attach it to the test leads, and try it out. When you are satisfied with its operation, slide it into the plastic case and the job is done. Be sure it works before putting it into the case. It may be hard to remove once it has been inserted.

There are several precautions that must be taken for the sake of safety. The range of this instrument is 1 millivolt to 1000 volts. Measurement of high voltages should not be undertaken unless one is aware of the potential danger. At voltages above 30 volts, human safety becomes very important. The entire circuit must be insulated from the metal cabinet to prevent the metal cabinet from becoming "hot" if the ground test lead were to be accidently attached to a high voltage point rather than to ground.

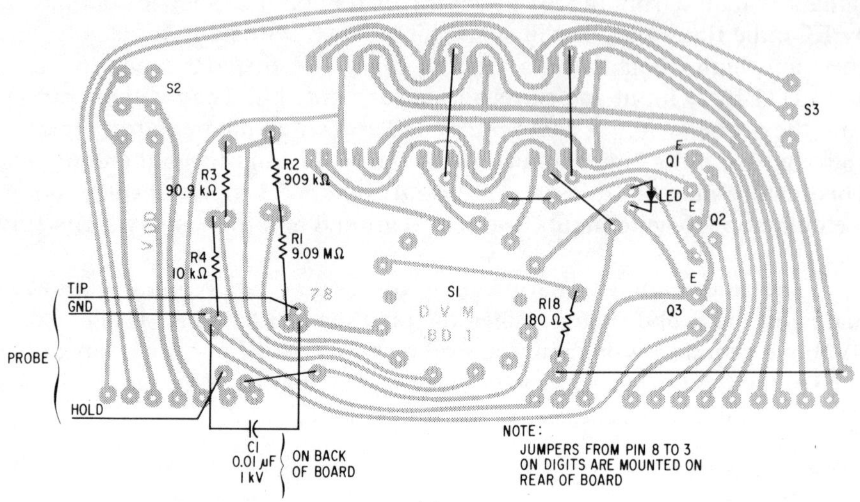

(A)

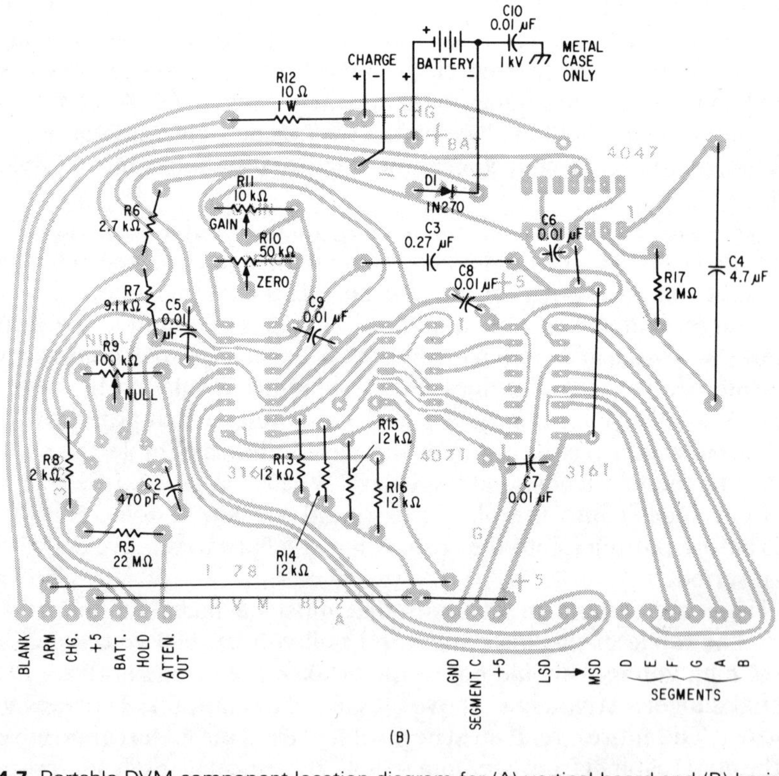

(B)

Fig. 4-7 Portable DVM component location diagram for (A) vertical board and (B) horizontal board.

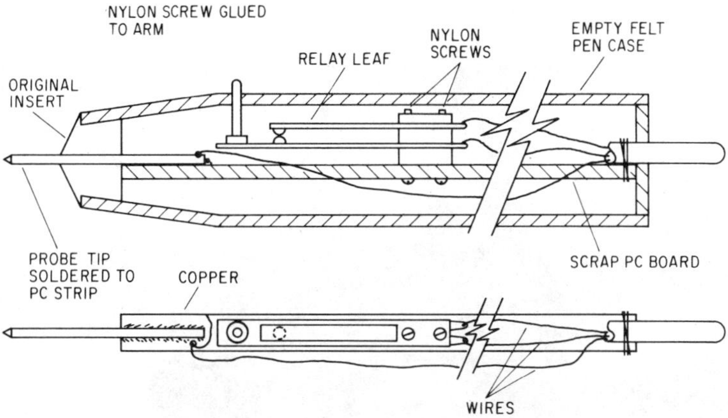

Fig. 4-8 Cross section of the DVM probe.

The only connection between the metal case and the DVM is the 0.01 μF, 1000 V capacitor C10 from the minus battery terminal. This is necessary to prevent stray ac pickup that can cause flickering of the digit. Here, the 1000-volt capacitor rating is essential. Likewise, the 0.01-μF disc capacitor across the test leads must also be rated at 1000 volts. If such high voltages are to be measured, be sure the test leads also have insulation that is rated at 1000 volts and, even then, test only with one hand in the pocket.

Insulation must also be provided between the charger jack and the cabinet. This is done by use of fiber shoulder washers between the mounting screws and the case. Also be sure to use a plastic probe case to insure against an accidental short between the probe case and a high voltage lead wire. An inside view of the portable DVM is shown in Fig. 4-9.

Calibration is an easy three-step procedure. First insert all the ICs except the CA3130. Be sure the locations of pin 1 are correct. With a short piece of wire, ground pin 6 of the CA3130 socket to the printed circuit board ground. With the range switch in the 1-volt position, adjust the zero trimmer to give a zero reading.

Second, remove the short, shut off power, and insert the CA3130. This will cause the zero reading to change. With the test leads shorted, adjust the null trimmer to again give a zero display.

Third, set the range to 10 volts and connect the test leads to a fresh, never used, 9-volt transistor battery observing proper polarity. Adjust the gain trimmer for a reading of 9.74 volts. The calibration is done. Of course, access to a precision voltage source or comparison with a precision DVM would allow for closer calibration than was obtained above. But is it really necessary to be better than 1 or 2% for the hobbyist?

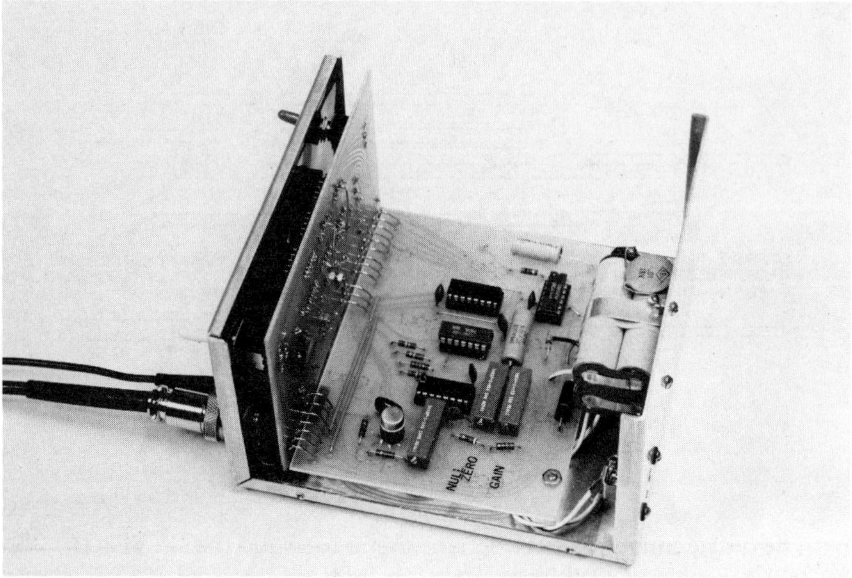

Fig. 4-9 Inside view of the portable DVM.

Battery Saver

Modern integrated circuits, with their low power drain, have made portable battery operated instruments very handy and practical. A good example is the digital voltmeter previously described.

DVM battery life has been extended by blanking the digits when they are not in use. Even so, if the power switch is left on overnight, the batteries will be dead by sunrise. What is needed is a circuit that can be permanently attached to the battery, draws negligible current itself, yet is able to turn off an electronic switch between the battery and the DVM circuit.

For low quiescent current drain, no device is better than a CMOS digital integrated circuit. This project describes such a switch (Fig. 4-10) and tells how to attach it to any battery operated package.

Circuit Description

This circuit (Fig. 4-11) measures a definite interval of time between turn-on of the major circuit (the load) and automatic turn-off. In the OFF mode, the total battery drain is on the order of 1 or 2 microamperes.

A small AA size nickel–cadmium battery has a life expectancy of 500 milliampere-hours. That means the product of current and time during use

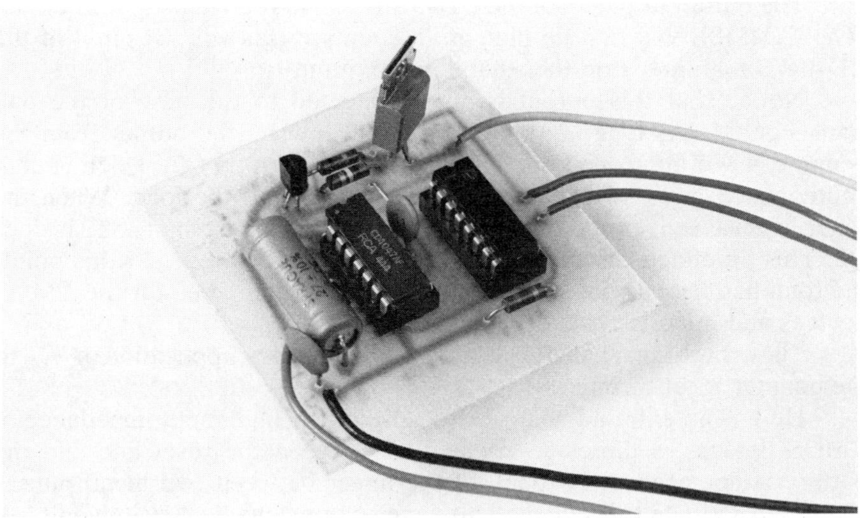

Fig. 4-10 Battery saver finished assembly.

is equal to 500. If the current is 50 milliamperes, the battery will run for 10 hours or, at 100 milliamperes, 5 hours and so on. As a worst case, assume the circuit draws 10 microamperes (0.010 milliamperes) in standby:

$$\frac{500 \text{ mA-hours}}{0.010 \text{ mA}} = 50{,}000 \text{ hours} \approx 5.7 \text{ years}$$

Obviously, the battery saver can be permanently connected to the battery with no effect on battery life.

Here is how it works. When the reset switch S1 is closed, the output from the CD4520B (pin 14) dual divide-by-16 counter will be reset low and will pull pin 4 of the CD4047B low. When pin 4 goes low, it allows the CD4047 to operate as an astable multivibrator. The frequency is set by R1-C1 attached to pins 1, 2, and 3.

The choice of time before shutdown is largely up to the user. For a test instrument that is used intermittently, such as the DVM, 15 minutes should be good. For this long a time, the CD4047B alone would require a capacitor much too large to be practical. This obstacle is easily overcome by using the CD4520B. This device is very similar to the CD4518B used in several other projects. Its only difference is that it divides by 16 in each section instead of by 10, for a total division of 256. This allows the CD4047B to be run at a higher speed with a resultant smaller value of C1.

The output of pin 14 of the CD4520B will first go high at 120 counts. (The CD4518B will first go high at 78 counts.) This will set pin 4 of the CD4047B high and stop the astable multivibrator.

Notice that this output is also connected to the base of the pnp transistor Q1. As long as this base is held low by the output from the counter, it will be turned on. This in turn will turn on Q2, which is connected as a series switch between the battery and the load. When the timer interval is over and the base of Q1 is driven high it shuts off, as does Q2. This, in effect, disconnects the load from the battery. At this point, the total battery current will be only a few microamperes for the CMOS devices and minor transistor leakage.

The switch can again be closed by a momentary application of V_{DD} to the counter reset through S1.

Here is an extra advantage. Because of the high input impedance of CMOS devices, a direct dc connection between the reset line and the battery is not needed. A touch of the finger between two metal pins is enough to activate the switch. This saves one mechanical switch and is a novel way to turn on an instrument.

The series transistor Q2 and its driver Q1 have an advantage over a conventional Darlington pair as the switch. The Darlington circuit, while simple, will cause a loss of about 0.7 to 0.8 volt between the battery and its load. This is due to the inevitable V_{BE} drop across the power transistor.

The circuit used in this project will only lose 0.1 to 0.2 volt, the saturation voltage of Q2. Thus, its use will not degrade the operation of the circuit it controls.

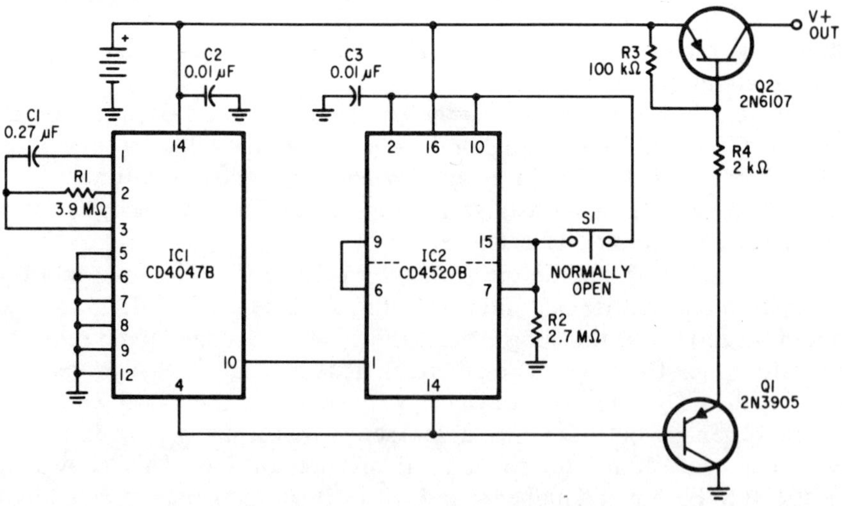

Fig. 4-11 Battery saver schematic diagram.

Parts List

C1—0.27-μF, 80-volt tubular C2, C3—0.01-μF, 100-volt ceramic
IC1—4047B monostable/astable multivibrator
IC2—4520B dual synchronous binary divide-by-16 counter
Q1—2N3905 pnp signal transistor Q2—2N6107 pnp power transistor
R1—3.9 megohms R2—2.7 megohms
R3—100 kilohms R4—2 kilohms
S1—SPST normally open momentary switch
Misc.—Printed circuit board, IC sockets, etc.

Construction

The entire circuit is mounted on a 2 in. × 2 in. printed circuit board. Fig. 4-12 shows the locations of the components. The board is intended to be added inside an existing battery operated instrument.

From the schematic (Fig. 4-11), we see two leads permanently attached to the battery. These are the ground and $+V_{IN}$. The $+V_{OUT}$ is connected to the circuit to be controlled and is labeled V_{OUT}.

The two wires from the side go to a switch for "turn-on" or just to two insulated contacts on the front panel for touch control. Only a skeptical hobbyist would worry about a mechanical switch being needed. However, one can be added if desired, right at the battery terminal.

As an example, this battery saver can be added to the DVM project (Fig. 4-13) or the frequency counter project. In each case, the line on the printed circuit board labeled V_{DD} requires a jumper to tie it into the circuit proper. To add the battery saver, remove the jumper. Connect the V_{IN} terminal to the solder dot on the line that runs directly to the power switch. Connect the V_{OUT} terminal to the remaining solder dot. All that is left is to connect the ground terminal to the negative side of the battery.

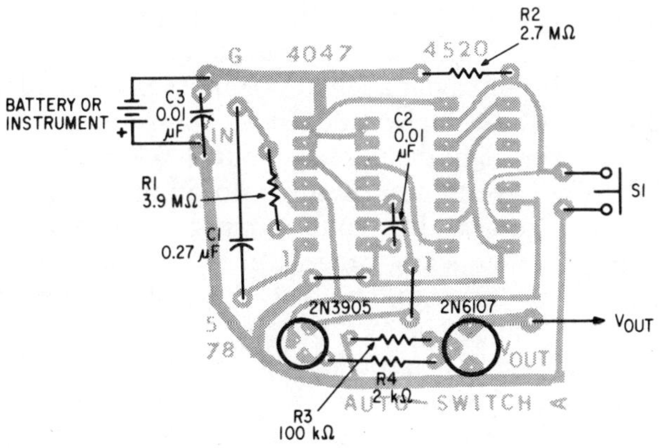

Fig. 4-12 Battery saver component location diagram.

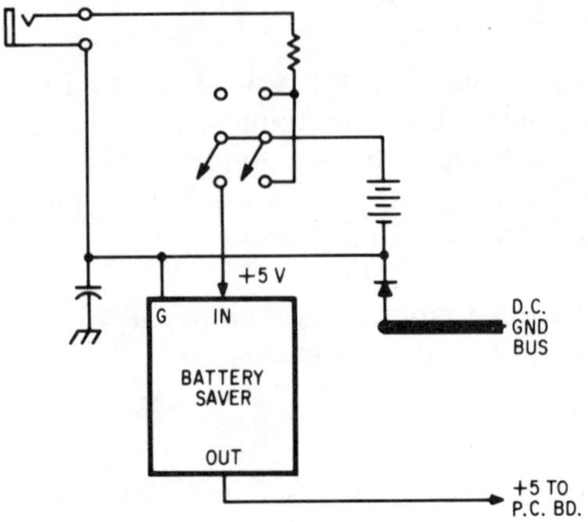

Fig. 4-13 Installation of the battery in the portable DVM. It is inserted in the $+V_{DD}$ line of the circuit.

The power switch of the DVM or frequency meter may now be left continuously on, or it might be turned off if the instrument is not to be used for long periods.

The operating time before automatic cutoff can be increased by use of a larger value of C1, or it can be reduced with a smaller value. R1 may also be reduced to shorten the time, but it should not be increased much above the 3.9 megohms specified.

In only a few minutes, your instruments have an extended battery life with no "Did I turn it off?" problem. Try it and see.

Battery Operated Frequency Counter

Many different types of frequency counters have been described in recent literature. Most of the articles emphasize the large number of digits that can be seen. Others tout the very high frequencies that can be read. Too often they don't talk about accuracy. None ask the reader if he really needs so many digits or so high a frequency. Think about it for a moment.

If a counter with eight digits reads up to 100 MHz to the last 1 hertz digit, does it really have that accuracy? More important, does the hobbyist need such accuracy? The large number of digits may only serve aesthetic purposes. Frequency range can also be an over-emphasized characteristic. For the hi-fi fan, one or two megahertz more than covers the best hi-fi audio equipment. A CB'er or Ham might want to reach 30 MHz to read his transmitter frequency on any band up to 10 meters.

The frequency counter to be described here (shown in Fig. 4-14) will cover 4.5 MHz by itself and will read up to 35 MHz or more when used with the divide-by-10 pre-scaler described as the next project. Although it

Fig. 4-14 Frequency counter finished assembly.

has only four digits, it has a dual range of 10 kHz and 10 MHz. The maximum upper range is limited by the CMOS devices. It uses a 0.01% accurate, 10 kHz crystal which allows for 1 hertz accuracy ±1 digit in the 10 kHz range. However, it does have the advantages of low battery drain, portable size, 0.01% accuracy, ± one count, and ease of construction.

Circuit Description

The heart of the project circuit (Fig. 4-15) is IC9, a 74C925 four stage divide-by-10 counter. The output is multiplexed to save on components and wire lines. Multiplex simply means that each digit has its readout data sampled and displayed on a periodic basis. In this case, it is 1000 times a second. What actually happens is that all the digits have their corresponding segments tied together. Each digit is turned on or off by its own single transistor that is synchronized to the available data. When segment data for the first digit is available, it also appears on the other three digits. But only the transistor that activates the first digit is turned on, so only it displays the data. In a like manner, each digit is activated in step with its data. The change is so fast the eye perceives it as all digits lit continuously.

Why all this fuss? First, with only one digit on at any given time, the power drain is greatly reduced. This makes battery operation even more enticing. Second, the complex wiring required to connect seven-segment lines from each of four digits to the counter is reduced to only one set of lines.

Parts List

C1, C5—0.2-μF, 20-volt ceramic
C2—0.47-μF, 100-volt ceramic
C3, C6—100-pF, 100-volt ceramic
C4—0.05-μF, 100-volt ceramic
C7 through C15—0.01-μF, 100-volt ceramic
I1—HP 5082-7405 display
IC1—4069B hex inverter
IC2—4518 dual synchronous BCD divide-by-10 counter
IC3—4520 dual synchronous binary divide-by-16 counter
IC4—4011A quad 2-input NAND gate
IC5—4073A triple 3-input AND gate
IC6—4081B quad 2-input AND gate
IC7—4047A monostable/astable multivibrator
IC8—4050B hex noninverting buffer and TTL driver
IC9—74C925 4-digit counter with 7-segment
J1—Charger jack
J2—Coax jack
Q1—RCA 40673 dual insulated-gate FET (or SK-3065)
Q2 through Q5—2N3241
R1—10 megohms
R2, R3—470 kilohms
R4, R6, R7—1 megohm
R5—100 ohms
R8—200 ohms
R9—750 ohms
R10—330 ohms
R11—1.2 megohms
R12—10 kilohms
R13—10 ohms, 1 watt
R14 through R20—820 ohms
S1—SPST switch (Alco MTB-106 or equivalent)
S2—DPTT switch, center off (Alco MTB-206P or equivalent)
XTAL—Statek SX-1H 10-kHz crystal (available from Godbout Electronics, Box 2355, Oakland Airport, CA 94614)
Misc.—Printed circuit board, IC sockets, suitable enclosure, etc.

The remainder of the CMOS circuitry provides the timing pulses necessary to operate the timer as well as the input amplifier to increase the instrument's sensitivity.

A CD4069B is used as a crystal oscillator at 10 kilohertz. This frequency is divided by 10 three times by the CD4018B dual divide-by-10 and half of the CD4520B. This last IC is really a divide-by-16 but has been converted to divide-by-10 by use of an AND gate (CD4081B; section A) to reset it on a count of 10. The second half of the CD4520B is used to

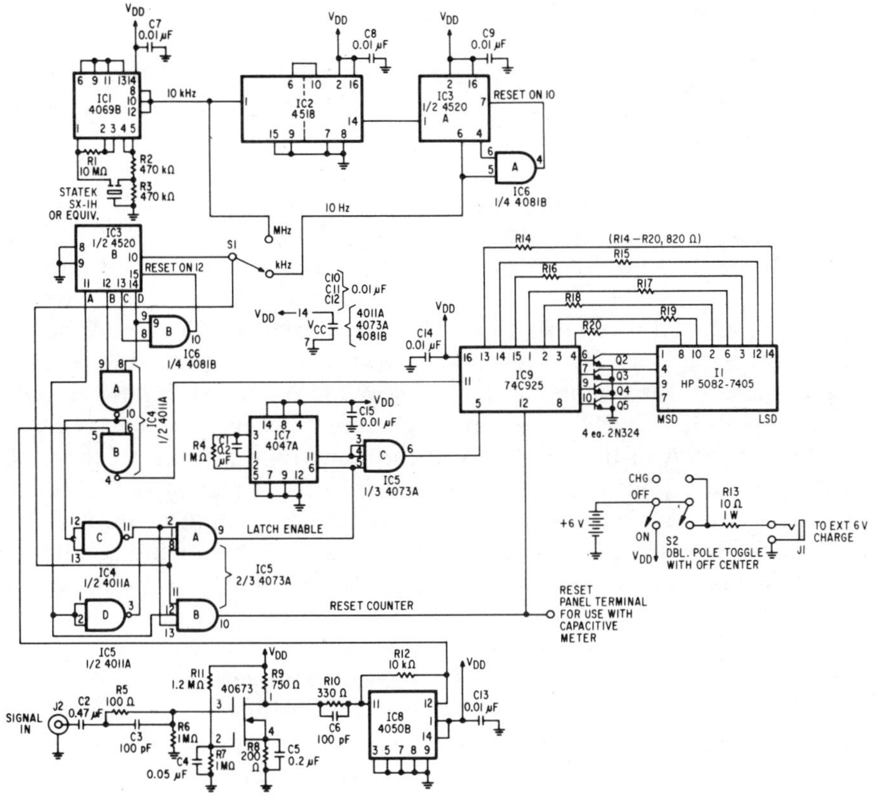

Fig. 4-15 Frequency counter schematic diagram.

provide the fourth divide-by-10 for a 1 second count time as well as two more pulses to update the visible digits and to reset the counter.

Let's start at time zero. The counter and the CD4520B are reset to 0000, the input signal has a path to the counter through section B of the CD4011A, the display is 0000, and the switch is set to read kilohertz. The four stage counter starts counting the signal pulses. This is how the timing logic works.

The CD4520B IC3 (section B) has four output lines A, B, C, and D that follow the changing binary count number. At the end of 10 counts (zero to nine inclusive), the output will read binary 1010 with the B and D lines high. These two lines are connected to the CD4011A (section A) NAND gate whose output is connected to a second NAND gate (section B) through which the incoming signal must pass. When the B and D lines are high (logic 1) the first NAND gate output goes low. This, in turn, drives the second gate off and stops incoming signals from reaching the counter. What has happened is that the counter has been open to receive signals for one second. The number of counts it receives is, therefore, counts per second or hertz.

Now that the four stage counter is on standby, a latch-enable pulse must be used to move this count from its internal memory to its external lines for display. This method prevents display flicker while the count process is in operation. The latch-enable pulse takes place immediately after the signal gate has been turned off.

Assume for the moment that the AND gate (section C of the CD4073A) is set so pulses can feed through it. Section D of the CD4011A is wired as an inverter. With line A from the CD4520B low, the inverter output is high and is fed to a CD4073A (three input AND gate). When all three inputs become high, the output will drive high and send a latch-enable pulse to the counters. Here is how the other two inputs go high. Section C of the CD4011 will have a high output because its input is driven low by section A of the CD4011A at the same time it turns off the input gate. The third input is connected to the 10 hertz signal at the CD4520B input. As soon as one high pulse is received, the output of the CD4073 (section A) will drive high and trigger the latch-enable of the counters.

Likewise, on the next count out of the CD4520B, a pulse will be provided to reset the counters and the cycle starts over again. It will be left to the reader to figure out this similar logic.

This last count also resets the CD4520B to zero by means of the CD4081A (section B) connected in the same manner as previously described for the third crystal frequency divide-by-10 stage. The system is now back to the point at which this discussion started.

From this it can be seen that a 1 second sample of incoming signal has been counted and moved from memory to display. The system has then been reset to repeat the process.

Only a single wire change is needed to provide the megahertz range. The input to the last half of the CD4520B is switched from the three divide-by-10 stages to the crystal frequency itself. This makes the count time interval 1 millisecond and in effect divides all megahertz readings by 1000. The display is now read as megahertz instead of kilohertz. The only problem here is that the display is changing 1000 times a second. Unless the input signal being measured is extremely stable, the last digit will flicker and be unreadable. This is overcome by use of a CD4047A timer added to the latch-enable line. Each time the negative-going side of a latch-enable pulse is received at the timer input, the timer output will be driven low and will remain there for a time interval determined by its RC time constant, about half a second in this case.

When the output is low, the AND gate of CD4073 (section C) to which it is attached will be forced low. This prevents any latch-enable pulses from reaching the counter which in turn prevents any change in the display. The reset and count functions of the counter are still in operation;

only the display is frozen. As a result, every second the display is sampled twice instead of 1000 times. The result is no flicker. In the kilohertz mode, this timer will have no effect since its output will be triggered low at the end of the first enable pulse and will return high before the next latch-enable command. With its output high, the following AND gate will pass the latch pulse and cause a display change once every 1.2 seconds.

Construction

The purpose of this project is to create a portable, battery operated, practical frequency counter. The need for a small package requires a fairly tight printed circuit configuration. The only critical part of the layout (Fig. 4-16) is the input preamplifier, where we must insure against unwanted feedback. Care must also be exercised when drilling the front panel. Be sure the switches align with the printed circuit board and panel holes before soldering them to the board. Don't apply any force to the board that might cause it to warp or crack. The digit readout may be moved closer to the panel opening by inserting one extra socket between the LED display and the soldered socket.

The battery charger must be insulated from the panel unless the panel is made from plastic. The most common chargers have the positive output connected to the frame side of the jack. Since the metal panel is connected to the ground line, a short circuit will appear across the battery through the 10-ohm charge resistor unless the jack is insulated. Nylon screws and washers are very effective here. If a plastic panel is used, be sure to add a ground wire from the coax jack outer shell to the ground bus on the printed circuit board. If the panel is metal, tie it to the printed circuit board ground line. The front-panel/board assembly is shown in Fig. 4-17.

An extra jack is shown on the panel. It is a wire connection to the reset line. It is used only to provide a reset pulse for use with the capacitance meter project. If not needed, it may be omitted.

Unlike all the CMOS devices discussed in previous projects, the 74C925 can only be operated with supply voltages from 3 to 6.5 volts. Higher voltages will destroy it. However, the higher the voltage, the higher the frequency response. Therefore, five rechargeable batteries are used to supply 6.2 volts for 4.5 megahertz operation. With only 5 volts, the frequency response will fall to 3.5 MHz.

There are no adjustments to be made. Just apply the power and signal to be measured. Start in the kilohertz range to be sure everything is correctly assembled. At inputs above 9.999 kHz, the reading will flow out the left side of the display. Switching to the megahertz range will display the over-range on a scale that reads up to 9.999 MHz. For example, 6.543 kHz will appear just that way in the kilohertz range. When switched to megahertz, the reading will be 0006 ± 1 count. A reading of 6.543 MHz will so appear in the megahertz range, and the kilohertz range will show 3XXX. The left-most digit in kilohertz should agree within ± 1 digit with the right-most digit in the megahertz range. The XXX represent the last significant digits, but do not expect 1 digit accuracy.

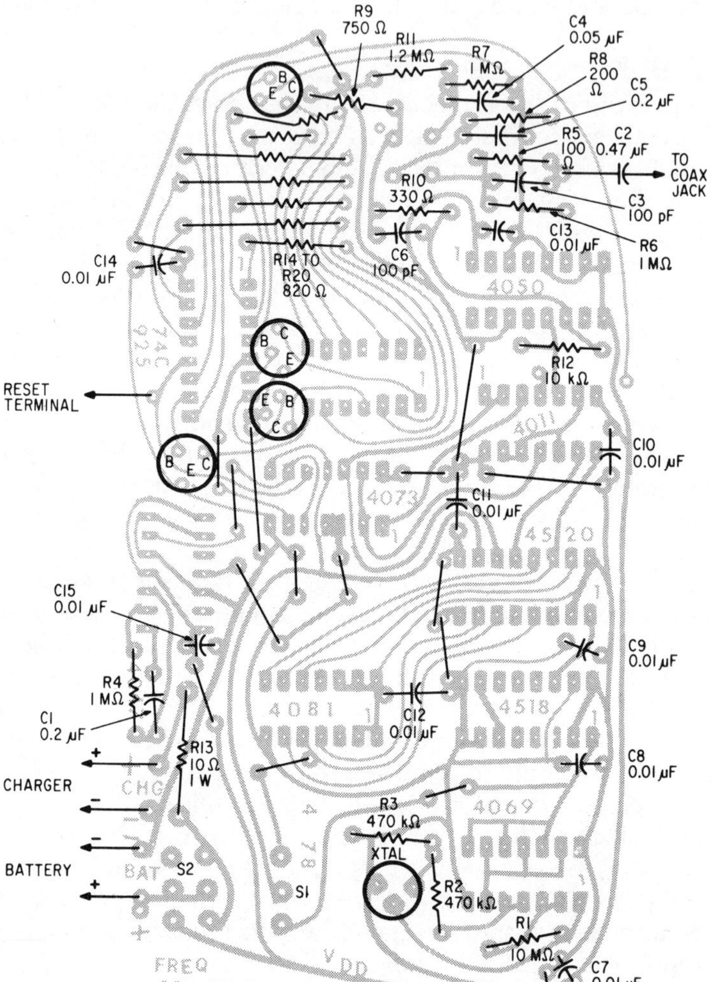

Fig. 4-16 Frequency counter component location diagram.

Have fun with your new counter. Use it with the divide-by-10 prescaler (next project) to extend its range or with the capacity meter to add to its usefulness.

Fig. 4-17 Side view of the frequency counter assembly.

Divide-by-10 Pre-Scaler

The very low current drain of the previous project (a portable frequency counter) was made possible by use of CMOS integrated circuits. The only disadvantage of CMOS devices is their frequency limitation of 4 to 5 MHz. Even in this range a counter still has many uses, as was described.

Added versatility can be obtained by using a divide-by-10 pre-scaler assembly (Fig. 4-18) to increase the range to about 45 MHz. To do this, it is necessary to use TTL logic. A proper selection of devices allows the coverage of all Ham bands to 10 meters, as well as the CB frequencies. The amateur radio operator who wishes to cover 50 or 144 MHz will need a different (ECL) divider.

Circuit Description

The input stage of this divide-by-10 circuit (Fig. 4-19) uses a 40673 DMOS amplifier in a circuit similar to the input stage of the counter. However, to insure adequate gain over the frequency range there are several changes.

An RF choke has been added to the drain load. This provides high frequency peaking and allows operation up to at least 45 MHz. A trim pot has been added so the load current can be adjusted to provide a drain voltage approximately half-way between the TTL high and low trigger voltages. The output of the amplifier is fed to the input of an inverter, the SN7404. Its purpose is to provide a square wave output to drive the divide-by-10 counter.

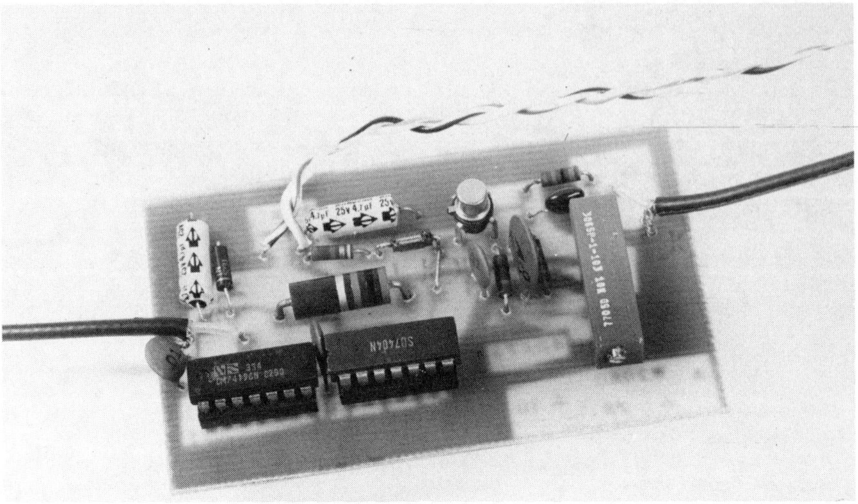

Fig. 4-18 Divide-by-10 pre-scaler finished assembly.

The SN74196 was chosen for the divide-by-10 because it is rated to accept inputs at least up to 50 MHz. Its output, therefore, is 5 MHz or less. This can now be fed into any frequency counter that is workable to 5 MHz. The portable frequency counter described previously is usable up to 4.5 MHz.

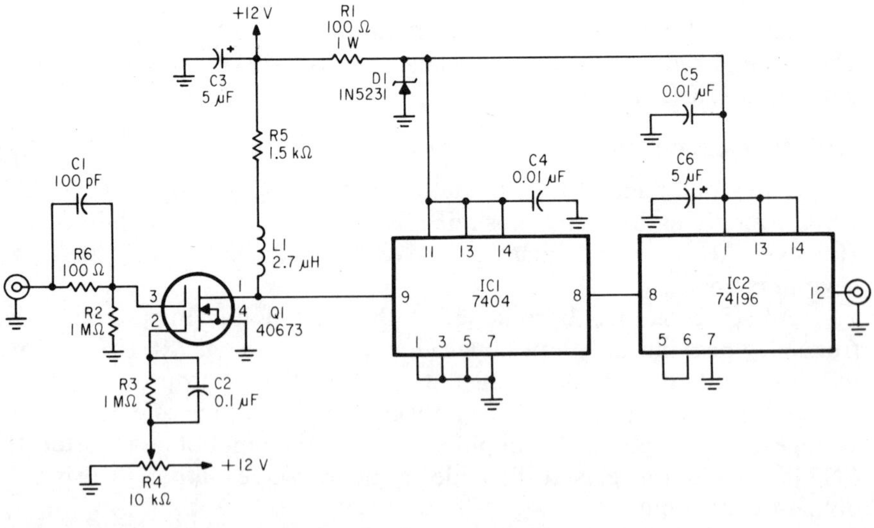

Fig. 4-19 Divide-by-10 pre-scaler schematic diagram.

Because the DMOS amplifier is operating with a 12-volt supply, it is necessary to add a dropping resistor R1 and a 5-volt Zener diode to provide the required 5-volt supply for the TTL devices. Adequate by-passing is provided by both electrolytic and ceramic capacitors. Certainly this would also be a good place to make use of the Regulated Voltage Divider described in Chap. 1.

Parts List

C1—100-pF mica
C2—0.1-μF, 16-volt ceramic
C3, C6—5-μF, 25-volt electrolytic
C4, C5—0.01-μF, 100-volt ceramic
D1—1N5231 Zener diode
IC1—7404 hex inverter
IC2—74196 decade divider
L1—2.7-μH choke (Miller 70F276A1 or equivalent)
Q1—RCA 40673 dual insulated-gate FET (or SK-3065)
R1—100 ohms, 1 watt
R2, R3—1 megohm
R4—10-kilohm trim-pot (Bourns 3068P or equivalent)
R5—1.5 kilohms
R6—100 ohms
Misc.—IC sockets, printed circuit board, RG174-U coaxial cable, etc.

Construction

The printed circuit board is self-explanatory. The only comment is that the extensive ground plane is used to reduce any chance of feedback oscillation at the higher frequencies.

When all the components are soldered in and the layout (Fig. 4-20) checked for errors, insert only the 40673 DMOS and apply the +12 volts. Adjust the trim pot to give approximately 1.4 volts on the 40673 drain (pin 1). Check the voltage at pin 14 of both the SN7404 and SN74196. It should be close to +5 volts. If it checks okay, insert the TTL devices. The input and output should be connected through small diameter coaxial cable such as RG174-U. Connect the output to a frequency counter and apply an RF signal to the input. Start at about 4 or 5 MHz. The frequency counter should read one-tenth of the input signal. If there is no reading, adjust the trim pot until a reading appears. Now increase the signal frequency. Should the counter stop indicating an output, retrim the pot.

There should be no trouble reaching 30 to 35 MHz, the upper limit resulting from use of the SN7404. Devices that operate up to 30 or 35 MHz are quite common. Operation at still higher frequencies will require trial-and-error testing of several devices. If trouble is encountered, first be sure the signal generator is putting out enough signal at the higher frequencies.

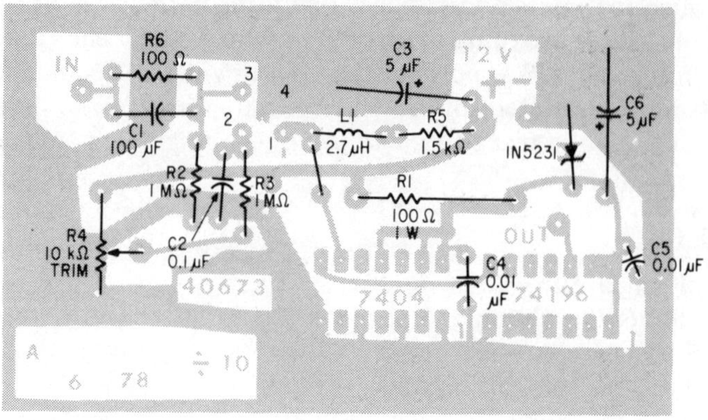

Fig. 4-20 Divide-by-10 pre-scaler component location diagram.

Many hobby signal generators do not have the required 0.3- to 0.4-volt signal at the top of their range. A useful test can be made by loosely coupling the divide-by-10 to a Ham transmitter whose frequency is known.

With this pre-scaler, the range of the portable frequency counter is extended to cover all Ham bands and shortwave bands up to 10 meters.

Capacitance Meter

By the time the Ham, hobbyist, or experimenter has built a few projects, he has an instrument to measure volts and ohms. This is usually enough. Occassionally, however, it is very helpful to be able to measure capacitance. An example would be the audio filters described earlier. While the exact capacity was not critical, matching of components was important. A laboratory capacitance bridge would be ideal, but such an expensive instrument is not often available to the home builder.

This project describes a small, inexpensive attachment to be used with almost any frequency counter. The capacitance meter (Fig. 4-21) will provide direct digital readout of capacitance on a frequency counter. The only addition to the counter itself is a small jack to provide access to the counter's reset bus. The frequency counter project already has this provision.

Circuit Description

Two NAND gates of a CD4011A are connected as an astable multivibrator, as shown in Fig. 4-22. The oscillation frequency is a direct function of the values of R1 plus R2 and C1. R2 allows frequency adjustment at calibration time.

Normally both inputs to each gate would be tied in pairs as inverters to provide free-running oscillation. In this circuit, one input is returned to the output of the CD4047A connected as a monostable multivibrator.

The time interval during which this multi is on is again a direct function of the values of resistance and capacitance, R3 plus R4 and CX.

CX is the unknown capacitor being measured. The time interval will vary directly with the value of CX. As CX increases, time increases.

Remember that the astable multi is oscillating at a fixed frequency of 10 kHz. The total number of pulses that it puts out is related to how long the timer allows it to oscillate. These pulses are counted by the frequency counter and appear on the readout. By proper choice of frequencies, capacitance may be read directly.

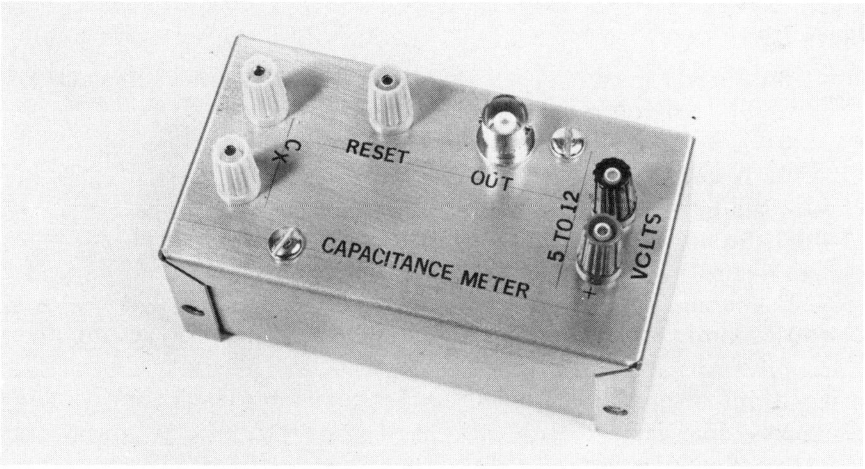

Fig. 4-21 Capacitance meter finished assembly.

Timing of the circuits to match the frequency counter in use is provided by the frequency counter reset line. At reset, the counter circuits are set to zero and the CD4047A timer turns on and allows the CD4011A oscillator to take off.

The oscillator will free run until it is turned off at a time interval directly related to the value of unknown capacitance. The number of pulses will appear on the counter and the job is done.

As designed, and when used with the frequency counter project, the unknown capacitor may be any value from 100 pF to 1 μF, the five-decade range most used.

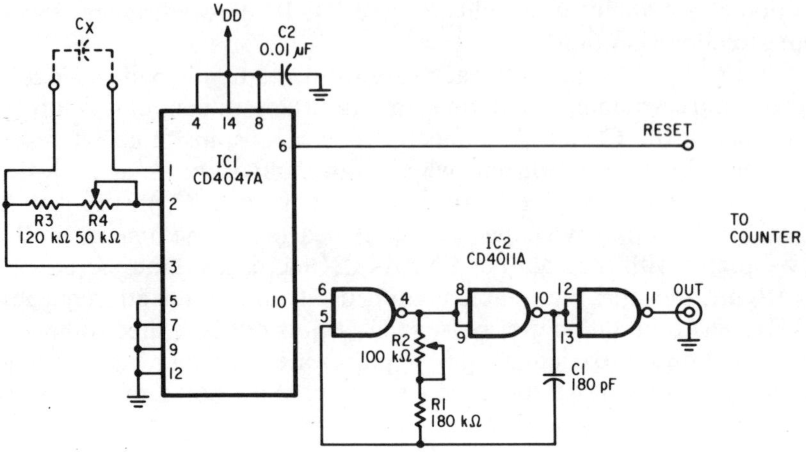

Fig. 4-22 Capacitance meter schematic diagram.

Parts List

C1—180-pF silver mica (Cornell-Dubilier CD10-CM04 or equivalent)
C2—0.01-μF, 100-volt ceramic
IC1—4047A monostable/astable multivibrator
IC2—4011A quad 2-input NAND gate
R1—180 kilohms
R2—100-kilohm trim-pot (CTS X201R-104B, Mallory MTC-54L1, or equivalent)
R3—120 kilohms
R4—50-kilohm trim-pot (CTS X201R-503B, Mallory MTC-15L1, or equivalent)
Misc.—Printed circuit board, IC sockets, suitable enclosure, binding posts, coax jack, etc.

Construction

Layout a small board as shown in Fig. 4-23 and insert all the components except the CD4047. The meter is now ready for calibration. Two 5% or better capacitors will be needed. Values around 0.47 μF and 0.0047 μF (4700 pF) are best, although any values in these ranges are good.

First apply between 5 and 12 volts and attach the frequency meter to the output terminal. The oscillator will be free running with its frequency appearing on the counter readout. Adjust R2 to give a frequency of 10 kHz. If the trim pot does not cover a wide enough range, vary the fixed resistor R1 up or down so that the trimmer can be used at about midrange.

Before inserting the CD4047A, be sure there is no short across the CX terminals. An accidental short here will blow the IC.

With the CD4047A in place, and with the 5%, 0.47 μF capacitor attached to the CX terminals, run a short lead from the reset terminal to

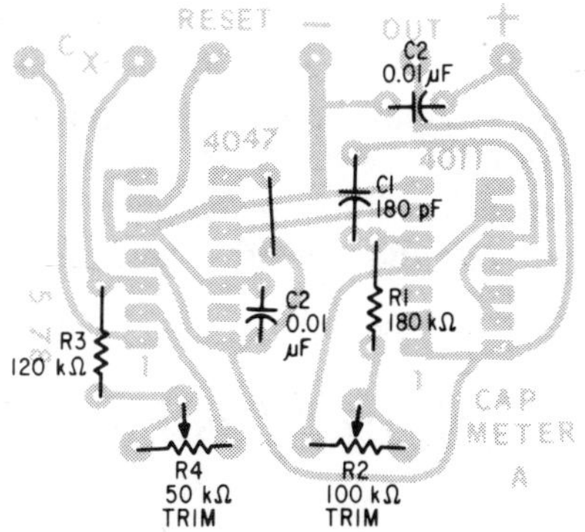

Fig. 4-23 Capacitance meter component location diagram.

the frequency counter reset line. With the power on, a reading should appear on the counter. The reading will be in microfarads and, therefore, the 0.47 μF should read 4700 with an imaginary decimal point at the left. Of course, the reading will not be correct until calibration is done.

Adjust R4 to obtain a reading of 4700. Now remove the 0.47 μF calibration capacitor and replace it with the 0.0047 μF one. The reading should be close to 0047. A fine adjustment of R2 on the astable oscillator may be necessary. Retry the 0.47 μF value. If it is within a few digits in the last place the job is done. An inside view of the capacitance meter enclosure appears in Fig. 4-24.

Fig. 4-24 Inside view of the capacitance meter.

If this meter is used with a frequency counter that has more digits, values up to 10 or 20 μF can be read with reasonable accuracy. Values below 0.0001 μF (100 pF) cannot be read and the accuracy below 0.0005 μF (500 pF) may fall off a bit.

Again, we see how a few inexpensive parts used with a previously built instrument can add one more useful tool to the work bench.

Digital Probe

There are many electronic tools common in industry that are not too well known to the hobbyist. One of these is the digital probe (Fig. 4-25). It allows one to study the state of any point in a digital circuit while the circuit is in operation.

A voltmeter could be used in the same way to determine if a logic point were high (logic 1) or low (logic 0). Actually there is no need to read the voltage. It is only necessary to know if the point is high or low. Why not use a single IC and some LEDs to give this information? Actually, the LED probe has several advantages over a voltmeter or even a DVM.

First, the LEDs are at the probe tip and are always visible. No need to move the eyes up and down between probe tip and meter. This means there is less chance for a slipped probe shorting an IC. Second, its response is faster. A moving-needle meter takes time to move and a DVM takes a second or two to stabilize before it can be read.

Third and most important, a voltmeter will not respond to a pulse, while this probe circuit, with its pulse stretcher, will. As the digital circuit under test changes state, the pulse detector will flash an indication of the change.

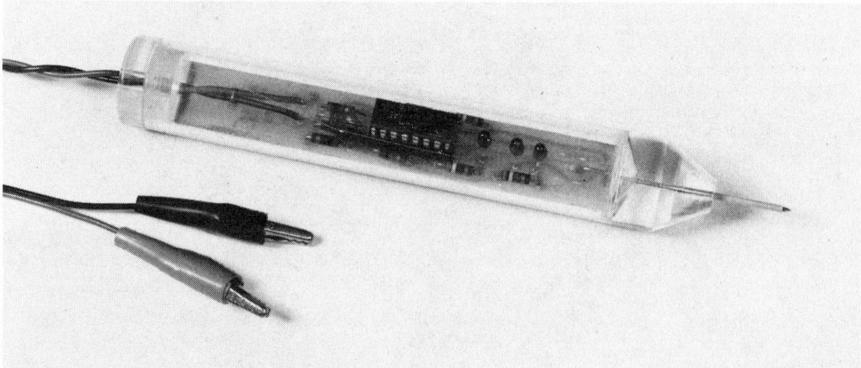

Fig. 4-25 Digital probe finished assembly.

Parts List

C1—100-pF mica
C2—0.02-μF, 100-volt ceramic
I1 through I3—Any subminiature LED
IC1—4049 hex inverting buffer and TTL driver
R1, R4, R8—1 megohm
R2, R6, R7—2.2 megohms
R3—20 megohms
R5, R9—10 megohms
R10 through R12—390 ohms
Misc.—Printed circuit board, alligator clips, suitable tube, etc.

Circuit Description

The probe is self-contained and receives its few milliamperes of operating current from the same power supply that feeds the circuit under test. The entire circuit (Fig. 4-26) is based on one CD4049A hex inverter integrated circuit. Since CMOS logic will operate on 5 to 15 volts, attaching the probe to the supply of the circuit under test allows it to be used

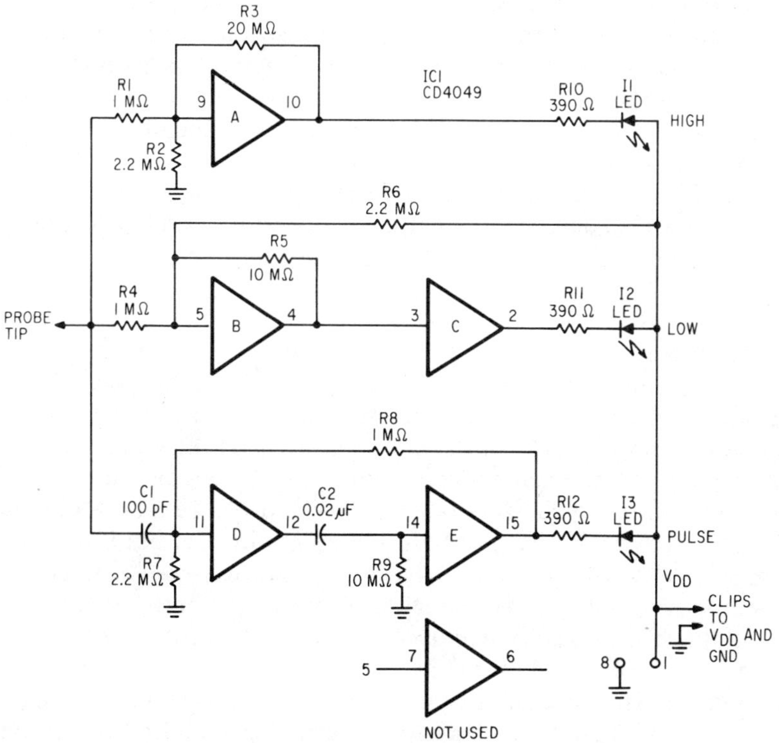

Fig. 4-26 Digital probe schematic diagram.

with TTL, CMOS, DTL, or any other type of logic in this voltage range. There is no fear of using the wrong power supply voltage. In other words, it is fail safe.

The probe tip is connected to the inputs of three independent circuits, one to indicate a high, one a low, and one to detect pulses. The CD4049A consists of six inverters. For any inverter, a high input will cause the output to go low and vice versa. Let's look at the "high" detector. When the input to inverter A is high, its output is low. This turns on the HIGH LED. Resistors R2 and R3 insure that this LED is off when no input signal is present. If the input is low, the inverter output will stay high and still keep the HIGH LED off. R1 isolates the inverter input from the other circuits and from the circuit under test.

The "low" detector operates in the same manner except that R5 the input to inverter B is returned to V_{DD} rather than to ground. Together with R6, this keeps the LOW LED off when there is no signal input. When the input goes low, the output of inverter B goes high. This in turn switches inverter C output low to turn on its LED. A high input will leave inverter C high and keep the LOW LED off. R4, again, is for isolation.

The pulse detector is only a little more complicated. It is connected as a pulse stretcher so that short pulses that would not give a visible flash of the LED are lengthened to about one-third second. This is long enough to make the flash easily visible. The length of the flash is set by the R9-C2 constant. Any pulse that occurs while the probe is in position will trigger the PULSE LED. C1 isolates the pulse detector from the dc input so that it only operates on a change of input voltage.

Since it is the negative-going side of a pulse that triggers the detector, a flash will also occur whenever the probe is touched to a low point or when a high point swings low, which is, after all, a pulse.

Construction

This project will require a little extra care and a steady hand. To make a fairly small, comfortable hand probe it is necessary to mount some components on each side of the printed circuit board. For this reason, examine the component layout sketches (Fig. 4-27). First, insert and solder all the components that go on the top of the printed circuit board. This will ensure that no solder dots become covered before a bottom component is soldered in. The IC can be soldered in directly, but if a very thin (low-profile) socket is available, its use will allow easy replacement of the IC, should it ever become necessary. Be sure all the parts lie close to the board so that the assembly will easily slip into the probe case.

The lucite case used here (Fig. 4-25) was made by drilling out a ¾ in. piece of lucite. The advantage of using transparent plastic is twofold. It is not necessary to align the LEDs with any openings in order to see them—

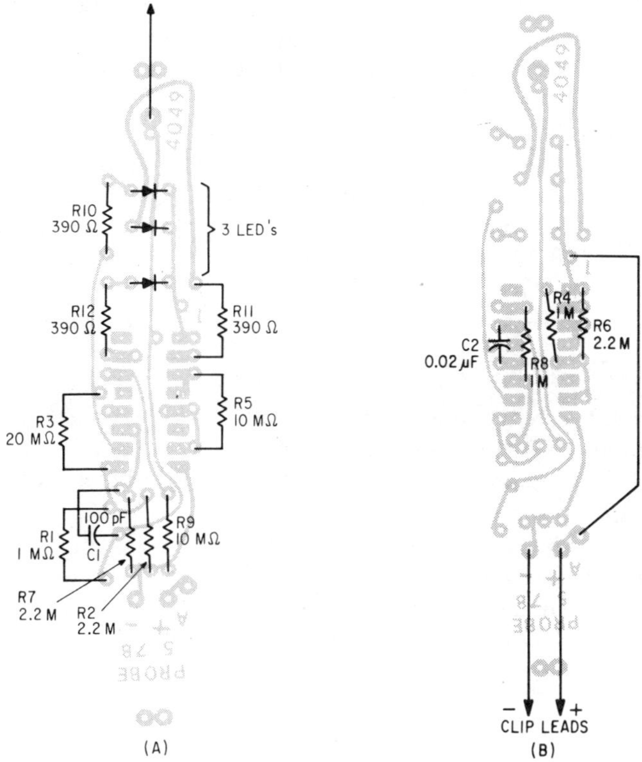

Fig. 4-27 Digital probe component location diagram for (A) the component side of the board and (B) foil side of the board.

they are visible from many directions. The transparent case also allows full visibility of the entire circuit. Show it to your interested friends.

If a nontransparent tube is used, it will be necessary to cut an opening in the tube to make the LEDs visible. Such an opening should be covered with a piece of celluloid or other transparent material to protect the probe circuit.

This digital probe will prove very useful in troubleshooting any of the digital circuits described in this book.

Logic Level Indicator

The digital probe described in the preceding project is a one-point device. It indicates the state at only one terminal at a time. Another useful tool is described here. With it, the state of every terminal of a digital IC can be seen simultaneously. Thus, if the input to a gate changes, the indicator will allow one to watch the output response. Another advantage of the multi-lead tester is that it seeks out the power leads on its own. Just clip it on and don't worry about where the plus and minus terminals are. The finished assembly is shown in Fig. 4-28.

Of course, every advantage may have a small disadvantage as a side effect. This circuit does not have a pulse detector. Its primary purpose is to troubleshoot slower circuits. It will show a terminal that does not respond as it should because of a bad IC, an open solder joint, or a short. It can also be used for education. As a deliberate change is made at one terminal, the results at all other terminals are instantly visible.

Parts List

D1 through D2—1N914 diode (32 required)
I1—Any small LED (16 required)
IC1 through IC3—4049A hex inverting buffer and TTL driver
R1—47 kilohms (16 required)
R2—680 ohms (16 required)
R3—1 megohm (2 required)
Misc.—Suitable enclosure and panel, IC sockets, 16-lead clip-on (Pomona Electronics "DIP CLIP" #3916 or equivalent), 16-lead ribbon cable, etc.

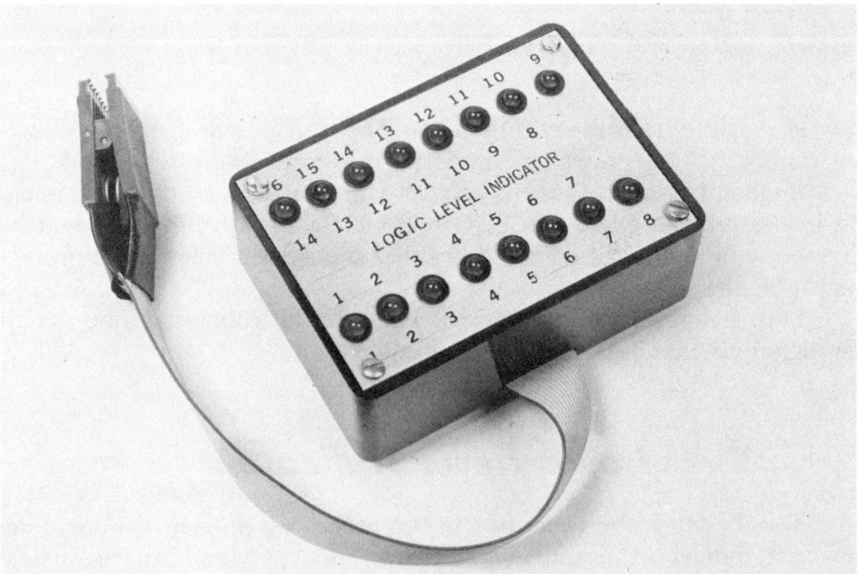

Fig. 4-28 Logic level indicator finished assembly.

Circuit Description

The same circuit is repeated sixteen times (Fig. 4-29), one for each pin of a 16 pin digital integrated circuit. In the case of 14 lead ICs, the unused two pins are left free but are protected as explained later.

Each pin is connected to the anode of one diode and the cathode of the other. If the pin is high or at V_{CC}, it is connected through D1 to a plus bus. If it is low, it is connected through D2 to the minus bus. Therefore, power for operation of the logic level indicator is obtained from the actual V_{CC} power pins of the IC under test as well as from all pins that are high. This provides the power for the inverters as shown by the connection of pins 1 and 8 of the CD4049A in the schematic. There are three CD4049A devices used to cover all 16 pins. Pins 1 and 8 are connected the same way for the other two devices.

In addition to supplying circuit current, each pin is connected through R1, a 47-kilohm resistor, to the input of an inverter. If the pin is high, the inverter output will be driven low and will turn on the corresponding LED. Resistor R2 limits the LED current to a safe value. If the pin is low, the inverter output will be high and the LED will be out.

When a 14-lead IC is being tested, the two unused contacts (pins 8 and 9 of a 16-lead IC) would be left floating. This is bad for CMOS ICs because the device current could rise and overheat the chip. Therefore,

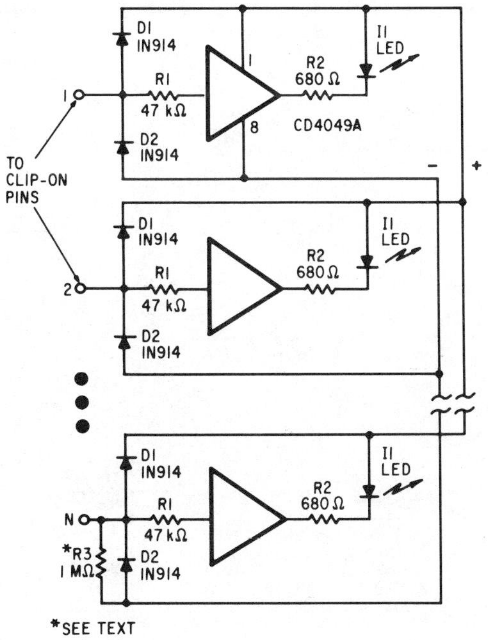

Fig. 4-29 Logic level indicator schematic diagram.

these pins are returned to the negative bus through 1-megohm resistors. With no input, the LEDs will stay out. When used on a 16-lead IC, operation will be normal. These 1-megohm resistors are not needed on the 14 remaining pins because they will always be making contact to the external IC under test. When removed from the circuit under test, there is no power applied and the inverters are off.

Construction

The layout for this printed circuit board is fairly tight to permit the entire assembly to fit into a small plastic case. The component location diagram appears in Fig. 4-30. A larger board can be used and will work as well. There are several construction hints that should be followed.

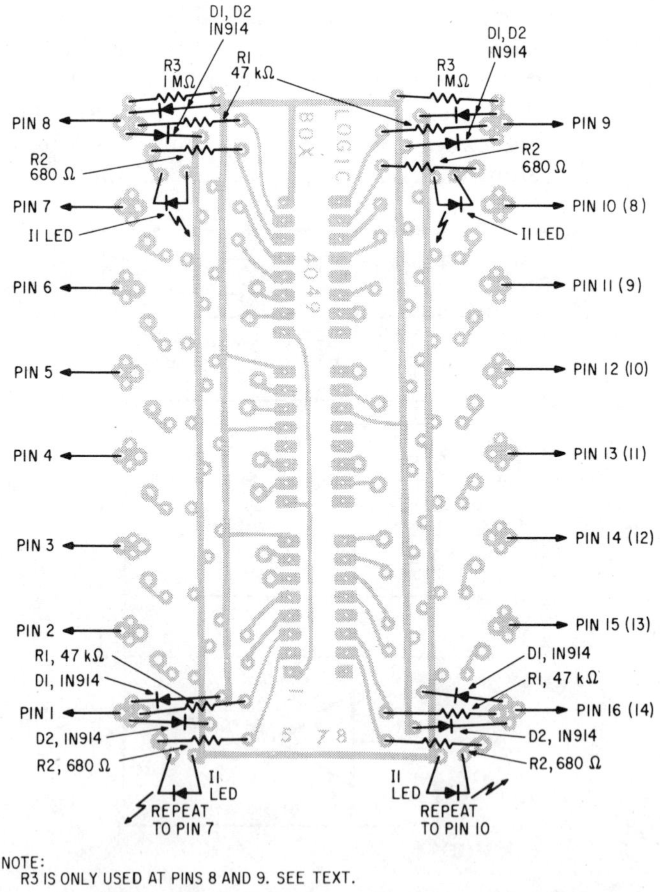

Fig. 4-30 Logic level indicator component location diagram.

First, solder in the IC sockets, the 680-ohm resistors, and the diodes. Be sure each diode is properly oriented, all anodes to the inner or negative bus and all cathodes to the outer or positive bus. Add the 1-megohm resistors to pins 8 and 9 only, as is shown in Fig. 4-30.

Next, take the metal panel and snap in the 16 LED holders. Except for the four corner LEDs, leave the holders loose. The reason will be obvious once you try assembling the whole structure. Tighten the holding rings only on the four corner mounts. After the printed circuit board is checked for good solder joints and proper wiring, slide the leads of each LED through their mounting holes—do it from the metal side of the printed circuit board. Be sure the anodes are connected to the positive bus. This will leave the IC sockets available for insertion and removal and will allow easier attachment to the external cable. Do not solder the LEDs yet.

Snap the four corner LEDs into their holders and adjust the space between the metal plate and printed circuit board to as wide as possible so that a soldering iron tip can be inserted. Be sure the assembly will fit into the case, with some room left at the bottom for the cable.

Now solder the four corner LEDs. This will make a rigid panel-to-board assembly. Take each LED and snap it into its mount. With the mount pressed close to the panel, solder the 12 remaining LEDs to the board.

A cable must next be prepared. Any 16-lead flat ribbon cable is ideal. Try not to use too thin a cable or there may be trouble working with its small-size wires. First cut about 1 to 1½ inches of each lead free at one end. Strip ¼ inch of insulation. This is for the clip-on. Separate about 3 inches of wire on the other end for the printed circuit board. Tin each lead at both ends. Also tin the pins on the clip-on. Slip a ½ inch length of shrink tubing on each of the 1-inch leads. Push it far enough back so it will not be heated when soldering.

Mark the outer wire on one edge of the cable with ink so that it can be identified as going to pin 1. Orient the clip-on to any IC so that pin 1 can be identified and marked on it. Solder wire 1 to clip-on pin 1. Skip one wire by pushing it to the other arm of the clip. Wire the next wire to pin 2. Continue using every other wire till all eight are soldered on one arm. Now solder the remaining eight leads in order to the other arm. Check that all solder joints are good and that there are no shorts between pins. Slip the shrink tubing down. Heat it carefully with the side of the soldering iron and the job is done. The fully wired board appears in Fig. 4-31. For added strength, a short piece of tape can be put around all the leads.

All that remains is to cut a slot in the plastic case just large enough to slip the ribbon cable through. Insert wire 1 in the solder hole for pin 1 and solder. Again skip one wire and solder the next one to pin 2 and so forth. When pins 1 to 8 are done, repeat for pins 9 to 16. If in doubt, check with an ohmmeter.

Fig. 4-31 Inside view of the logic level indicator.

The wiring can be easily checked by connecting 5 to 10 volts between the plus and minus buses. Then use a jumper from the plus supply to each input pin. The corresponding LED should light each time. If it does not, check for a poor solder joint. If the wrong LED lights, check for a misplaced wire pair.

For proof of its usefulness, attach the clip-on to any of the logic circuits described in this book and watch the lights show the logic levels as the circuit operates.

A very handy extra is to make a set of slip-on sketches of the most commonly used digital ICs to show the function of each pin for the IC under test. A template for the slip-on to match the LED spacing is shown (Fig. 4-32) together with a sample circuit (Fig. 4-33).

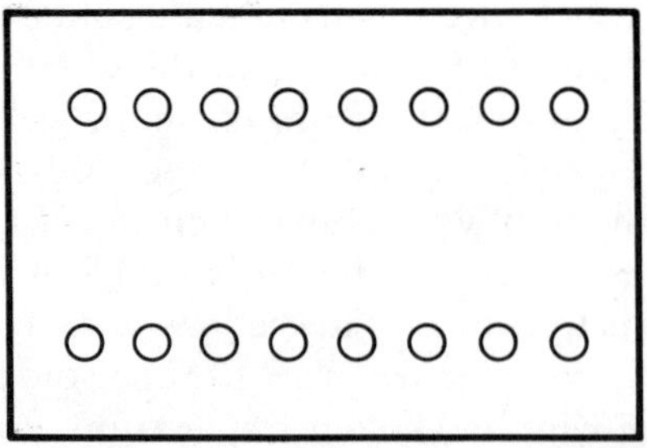

Fig. 4-32 Template for making information cards for the logic level indicator.

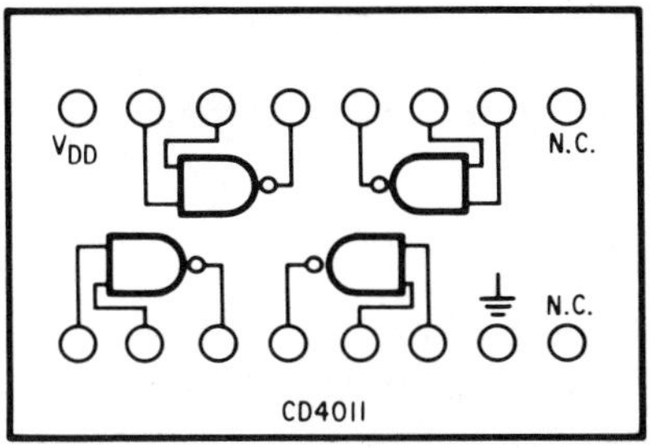

Fig. 4-33 Typical information card for the logic level indicator. This one shows the CD4011 quad 2-input NAND gate.

Dwell-Tachometer

In case you are afraid that a dwell meter-tachometer (dwell-tach) is a complicated piece of electronics, be relieved. The circuit is really quite simple and, in fact, half of it is a duplication of part of the DVM project.

This instrument (shown in Fig. 4-34) is most useful to those who like to tune up their cars or partake in auto rallies when knowing the engine

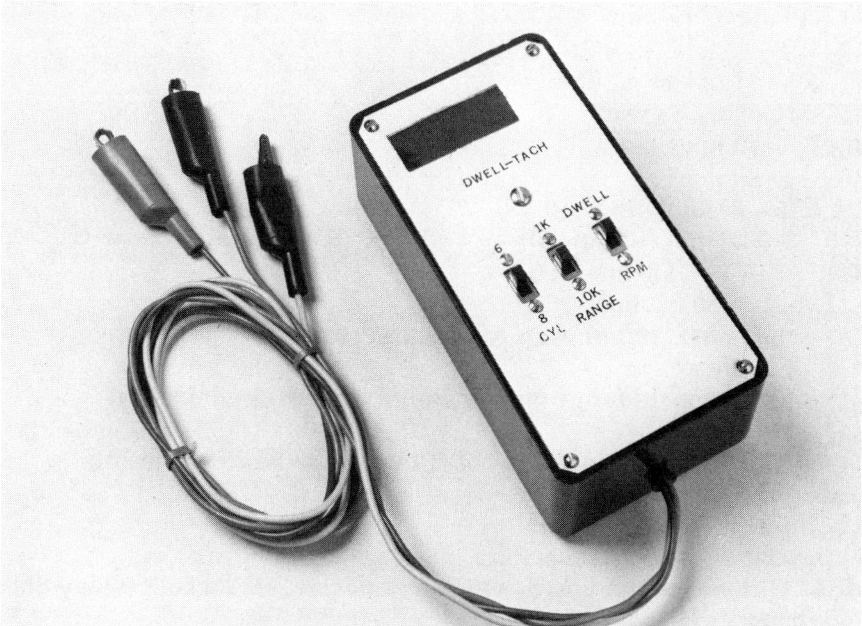

Fig. 4-34 Dwell-tach finished assembly.

speed helps in obtaining the best engine efficiency. It should be noted that this instrument will only work with cars that use points in the ignition system. It cannot be used on pure electronic ignition engines.

Parts List

C1—100-μF, 100-volt electrolytic
C2 through C4—0.47-μF, 200-volt mylar
C5—0.047-μF, 100-volt mylar
C6—0.0047-μF, 100-volt mylar
C7, C10—50-μF, 25-volt electrolytic
C8—5-μF, 25-volt electrolytic
C9—0.01-μF, 80-volt tubular
C11—0.33-μF, 80-volt tubular
C12—0.001-μF, 80-volt tubular
I1 through I3—FND 507 7-segment display
IC1—555 (or SK-3564)
IC2—RCA CA3161 BCD-to-7-segment decoder/driver
IC3—RCA CA3162 analog-to-digital converter
IC4—LM7805 5-volt constant voltage regulator
Q1—2N2102 npn power transistor
Q2 through Q4—2N4403 pnp transistor
R1—10 ohms, ½ watt
R2—1.5 kilohms, ½ watt
R3, R4—510 ohms
R5—510 ohms, ½ watt
R6, R7—10 kilohms
R8—120 kilohms
R9, R16—10 kilohms
R10—200-kilohm, 10-turn trim-pot (Bourns 3069P or equivalent)
R11, R20—49.9 kilohms, 1%
R12, R21—150 kilohms, 1%
R13—50-kilohm, 10-turn trim-pot (Bourns 3069P or equivalent)
R14—1 kilohm
R15—10-kilohm, 10-turn trim-pot (Bourns 3069P or equivalent)
R17—510 kilohms
R18—500-kilohm, 10-turn trim-pot (Bourns 3069P or equivalent)
R19—200 kilohms
R22 through R24—200 ohms
S1 through S3—DPDT miniature slide switch
Misc.—Printed circuit board, suitable enclosure, IC sockets, heavy duty battery clamps, etc.

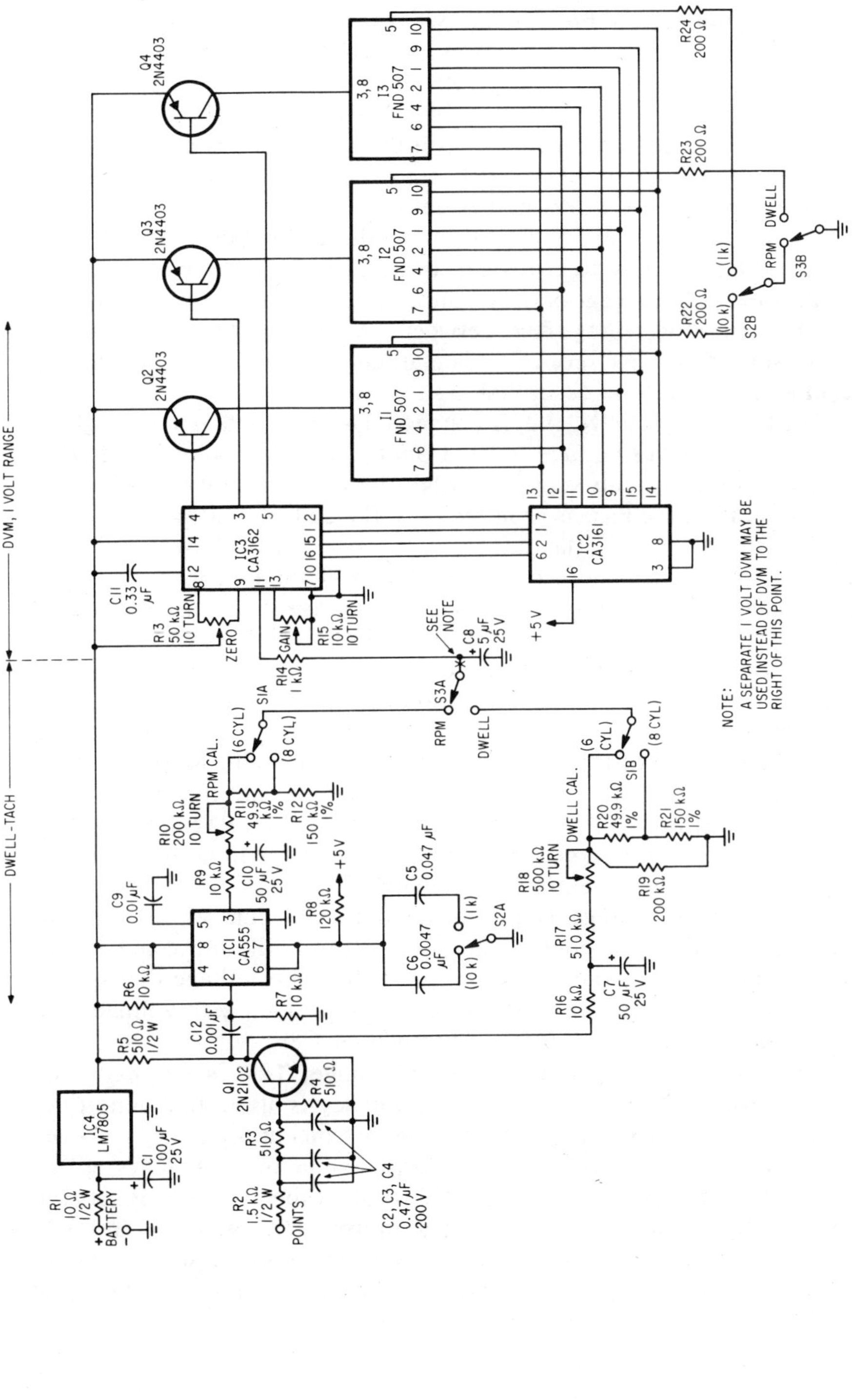

Fig. 4-35 Dwell-tach schematic diagram.

Circuit Description

This circuit may be broken into two distinct parts. The first is shown in the left half of Fig. 4-35 on the previous page. It consists of two channels, one for dwell time and one for RPM. The second part is a simplified single-range DVM. This part of the circuit has been described previously, so it will not be repeated here. In fact, if this dwell-tach is not to be used very often, the builder might choose to construct only the front-end and use a separate DVM for the readings.

There are two leads that connect to the car battery. An LM7805 voltage regulator is used to provide a constant 5 volts for circuit operation regardless of the actual car battery voltage under load. The signal input is taken from the low voltage line between the distributor points and the negative side of the spark coil. This is marked "points" on the schematic. The RPM circuit will be described first.

The base of Q1, a 2N2102, is connected to the signal line through an RC filter that removes voltage spikes from the line. The collector of Q1 is connected to the +5-volt supply through the load resistor R5. When the points are closed, the base is grounded and the collector sits at the 5 volts of the supply. As soon as the points open, the transistor base rises to the 12 volts of the car battery through the spark coil and its collector voltage drops to near zero. It is this fall in collector voltage that is used to trigger the CA555 connected as a one-shot multivibrator. Therefore, each time the points open, a pulse is generated to trigger the one shot. Each time it is triggered, a single pulse of fixed duration is produced. The faster the engine, the faster the pulse rate. The output pulse width is determined by the RC combination of R8 and either C5 or C6 depending on the range in use. These output pulses are averaged by R9-C10, and the resultant dc voltage is fed to the calibrated voltage divider R10, R11, R12. R11 and R12 are 1% resistors so that a single calibration pot R10 can be used to calibrate for both six- and eight-cylinder engines.

This average voltage is proportional to the number of one-shot pulses per second. The greater the engine speed, the more pulses per second and the greater the average voltage. It is this voltage that is read by the DVM. In other words, this circuit is a frequency-to-voltage converter. The DVM input is further smoothed by C8.

The DVM is set to a fixed one volt range. That is 999 millivolts equals 999 (or 9.99k) RPM. The pulse frequency is also a function of the number of engine cylinders. S1A sets the instrument for either a six- or eight-cylinder engine. S2A selects the correct value of capacitance, C5 or C6, to give the ranges of 1,000 or 10,000 RPM, respectively. S2B sets the proper decimal point for the RPM range in use. S3A switches the DVM input between RPM and DWELL. For the previous discussion, it would have been in the RPM position. S3B sets the decimal point when in the DWELL mode.

With S3A in the DWELL position, the DVM is now connected to an RC network. R16-C7 form an averaging circuit the same as R9-C10 in the RPM circuit. R17 through R21 are a calibrated divider to allow choice of six- or eight-cylinder operation. Here the average voltage is related to the duty cycle of the points, that is the percent of the time the points are closed. By definition, the number of degrees the points are closed is the dwell angle. S1B chooses six- or eight-cylinder operation.

As an example, in an eight-cylinder engine, each cylinder is entitled to one-eighth of 360° of the distributor rotation, or 45°. The points are closed for part of this 45°, known as the dwell time. For best engine operation, the manufacturer specifies dwell time or duty cycle in degrees. It is usually between 28° and 32°. With the dwell meter, the points can be adjusted to give the proper dwell.

Construction

Component location is shown in Fig. 4-36. All wire jumpers are located on the rear of the printed circuit board (Fig. 4-37). There are a few construction hints to be followed. The use of multi-turn trim-pots as shown is highly recommended for ease of adjustment. One percent resistors in the two dividers allow use of a single calibration pot for both six- and eight-cylinder engines.

In the absence of 1% resistors, careful selection of a pair having the proper values will help the overall accuracy. The important thing is to have a 1:3 ratio between R11 and R12 and between R20 and R21. The decade box and Wheatstone bridge can be useful here.

The panel is held to the printed circuit board by the three switch mounts. Note also that the voltage regulator is bolted to the aluminum panel to improve its heat dissipation. An inside view of the instrument is given in Fig. 4-38.

Use heavy lead wire about 4 feet in length with heavy clips on the ends to allow for rough handling and easy attachment to the battery terminals and points. Also, put rubber insulators over the clips since there can be high voltage transients present.

To calibrate the dwell-tach, a pulse generator and DVM, such as described previously, will be most useful. In the absence of these instruments, calibration can still be done as is explained later.

First, short pins 10 and 11 of the CA3162. Be careful not to short any other pins. Adjust the zero pot R13 for a zero reading. Remove the short. The switch positions will have no effect.

Second, set the pulse generator to give a 33 hertz, 50% duty cycle square wave of 12 volts from peak to ground and connect to the POINTS terminal.

With the DVM between pin 11 and ground, adjust the gain pot R15 to give the same dwell-tach reading as the DVM. Remove the DVM. This calibrates the dwell-tach DVM.

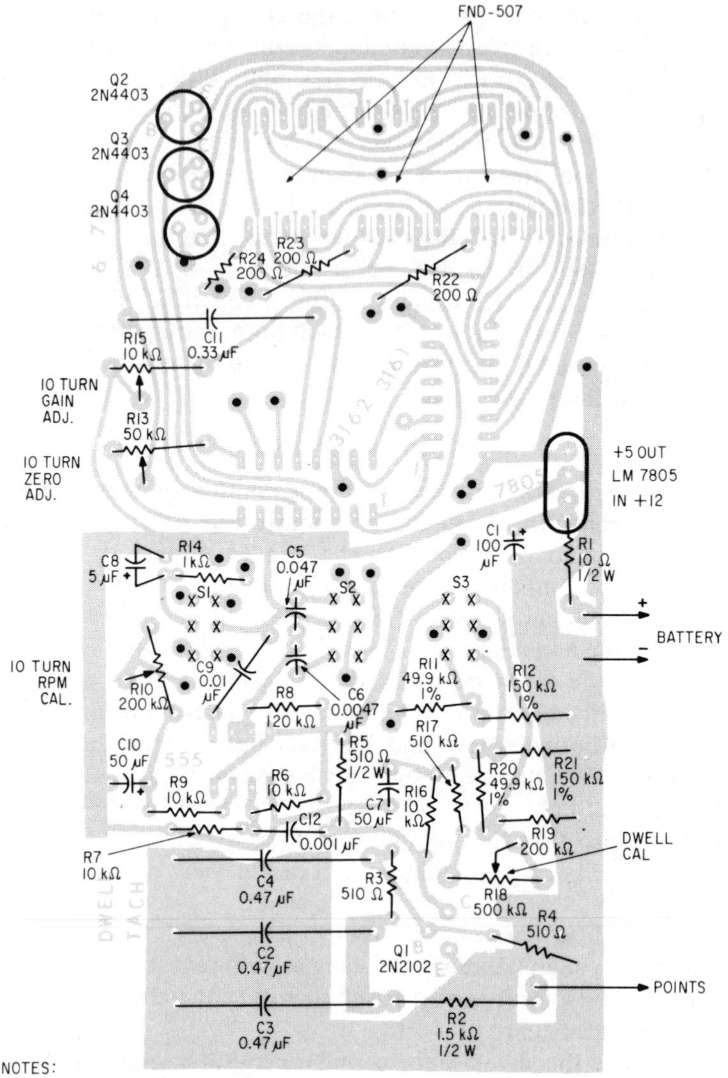

NOTES:
1. ALL WIRE JUMPERS ARE ON REAR
2. DOTS ARE JUMPER SOLDER POINTS
3. X's ARE SWITCH LEADS

Fig. 4-36 Dwell-tach component location diagram.

Third, set the switch to dwell and the cylinder switch to 8. Without touching anything else, adjust the dwell trimmer R18 to give a reading of 22.5°. Switch to 6 (cylinder) and the reading should be 30.0°. If the signal is removed, the readings will be 45° and 60°, respectively.

Fourth, set the switches to RPM and 8 cylinder and again without touching the input signal, adjust the RPM trimmer R10 to give a reading of 500 on the 1000 RPM scale.

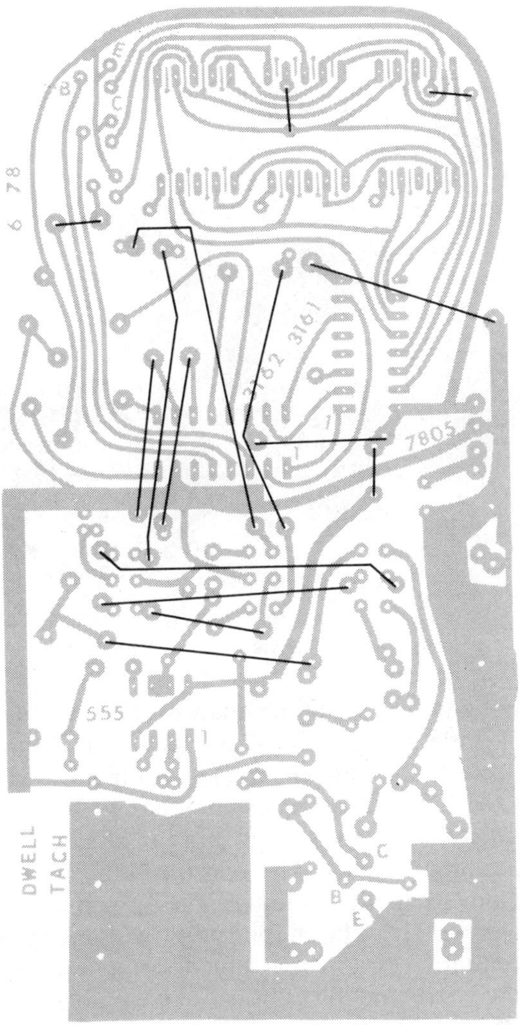

Fig. 4-37 Dwell-tach jumper location diagram. These are located on the foil side of the printed circuit board.

At this point, a final adjustment re-run through the calibration would be wise.

When switching between the 1k and 10k RPM ranges, there may be a slight disagreement depending on the accuracy of the capacity values C5 and C6 which set the one-shot pulse width. If it is desired to bring these readings closer together, a small capacitor may be paralleled to either C5 or C6 till the readings agree.

In the absence of a DVM and pulse generator, there is an alternate method for calibration.

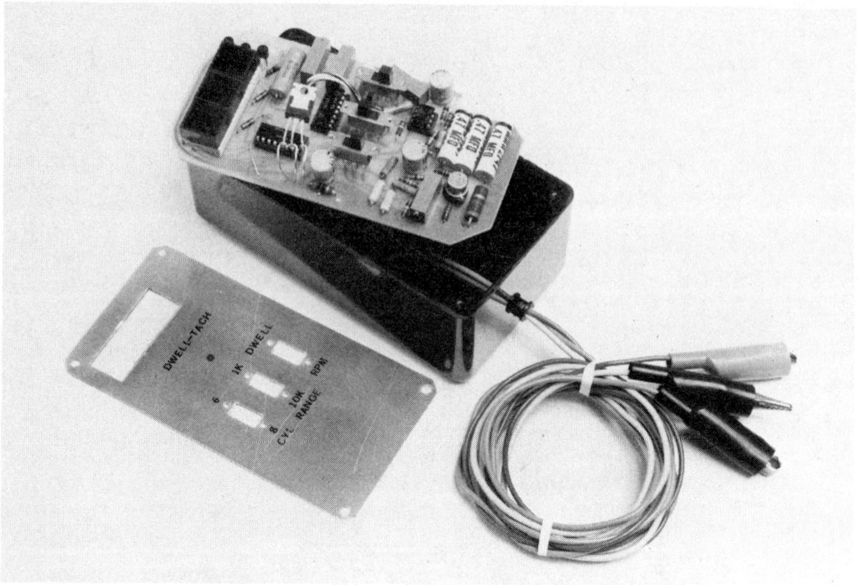

Fig. 4-38 Inside view of the dwell-tach.

A VTVM or any high input impedance voltmeter may be substituted for the DVM. A little more care will be needed in comparing the digital readout on the dwell-tach with the analog voltmeter.

To obtain the 33-hertz pulse, a small circuit will have to be wired. This circuit is very simple and is shown in Fig. 4-39. It is also used in the "Games" section that follows. The resistor value is adjusted to give 33 hertz by use of our frequency counter. The 12-volt amplitude is fixed by the 12-volt power supply. The output pulse will swing between 12 volts and ground. Calibration is now the same as originally described.

Now try the instrument on your car. An eight-cylinder engine should read between 28 and 32° dwell angle while six-cylinder cars should be between 38 and 42°. Check your auto manual for any manufacturer's recommendation.

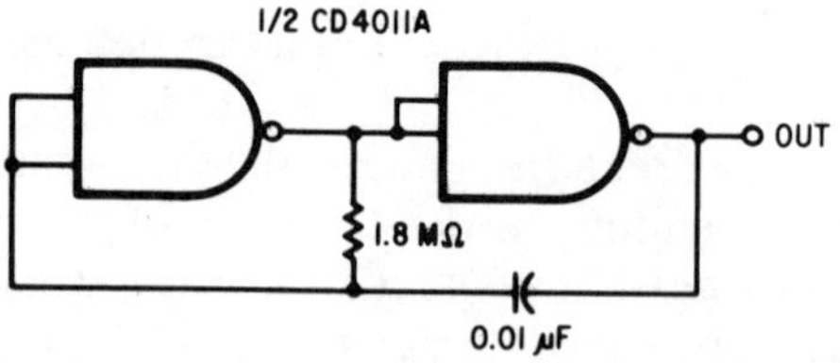

Fig. 4-39 Test oscillator circuit for calibration of the dwell-tach. R or C should be adjusted for 33 hertz output.

In the RPM mode, slow and fast idle speeds can be set. Again use the manufacturers specifications.

Only one precaution should be observed when using this meter. There are high voltage transients on the leads. Try to make electrical attachments with the engine off. Even a small shock, while harmless in itself, will cause a natural reflex. Remember, there are moving parts under the hood. They can become quite lethal.

5
Games

This book will end in a light-hearted mood, one of fun and gambling. The two games of chance described in this section were invented long before the discovery of the electron.

But electronics knows no bounds of time. Now, dice and roulette are played with lights and digits. With good gamesmanship, it might be possible to earn enough to pay for other construction projects! Have fun building and playing.

Electronic Dice

Anyone can buy a pair of white cubes and start playing craps. For novelty, however, an "electronic pair" is something that will make your friends envy your ingenuity and ability. The completed project is shown in Fig. 5-1.

Fig. 5-1 Electronic dice finished assembly.

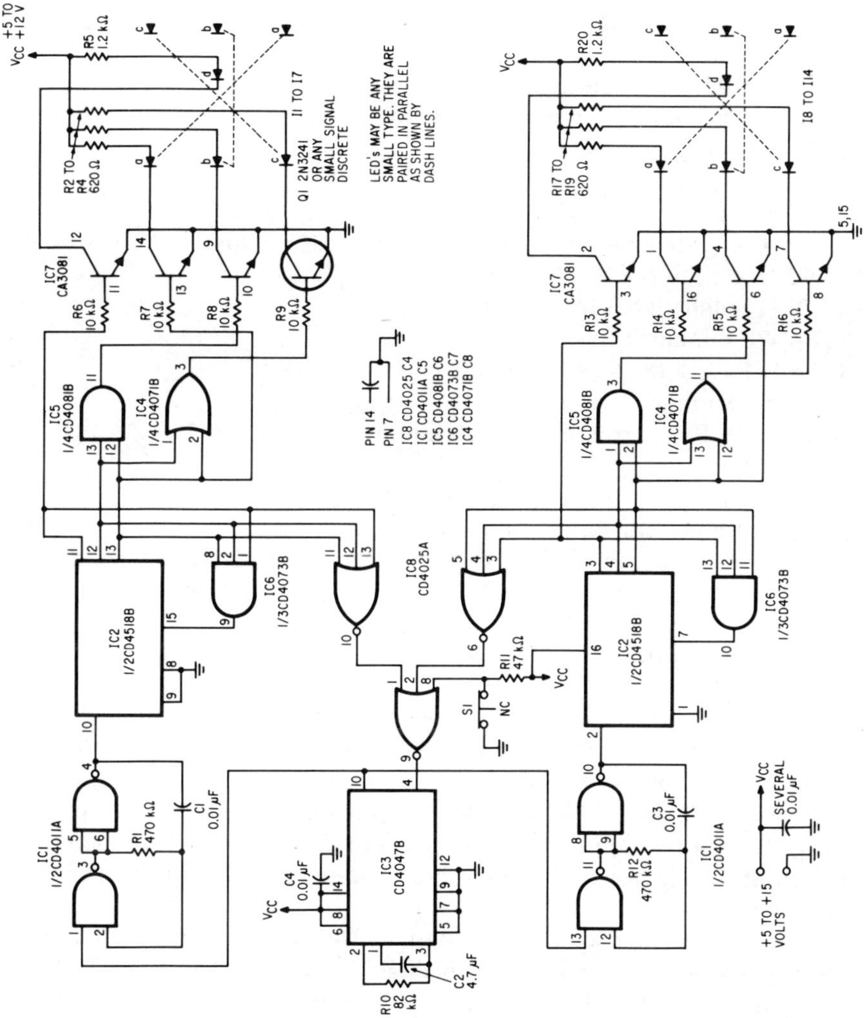

Fig. 5-2 Electronic dice schematic diagram.

Parts List

C1, C3 through C8—0.01-μF, 100-volt ceramic
C2—4.7-μF, 25 V nonpolar electrolytic (Radio Shack #272-998 or equivalent)
I1 through I14—Any miniature LED
IC1—4011A quad 2-input NAND gate
IC2—4518B dual synchronous BCD divide-by-10 counter
IC3—4047 monostable/astable multivibrator
IC4—4071B quad 2-input OR gate
IC5—4081B quad 2-input AND gate
IC6—4073B triple 3-input AND gate
IC7—RCA CA3081 high-current npn array (or SK-3550)
IC8—4025 triple 3-input NOR gate
Q1—2N3241 small-signal transistor
R1, R12—470 kilohms
R2, R3, R4, R17, R18, R19—620 ohms
R5, R20—1.2 kilohms
R6 through R9, R13 through R16—10 kilohms
R10—82 kilohms
R11—47 kilohms
S1—SPST normally closed momentary switch
Misc.—Printed circuit board, IC sockets, suitable enclosure, terminals, etc.

Circuit Description

As shown in Fig. 5-2 on the previous page, the schematic has a duplication of one circuit so that the pair can be "rolled" simultaneously. Only one side will be described.

A pair of CD4011 NAND gates are connected as a simple gated oscillator. One of the input lines is returned to a CD4047 timer. When the timer is on, the oscillator runs. After about one second, the timer turns the oscillator off.

The output of the oscillator is fed to a CD4518 divide-by-10 counter, whose output is in binary coded decimal (BCD) format. The CD4071 and CD4081 gates, as well as the direct divider output, form a "BCD-to-dice" converter. The CA3081 buffer transistors provide sufficient current to light the LEDs. The BCD that appears at the output of the divider is a function of the oscillator frequency and the length of time it is on. Random counts are ensured by the difference in frequency between the 1 second timer and the 500 cycle oscillator. In addition, neither oscillator or timer are given special treatment to insure their accuracy. In other words, oscillator drift is good.

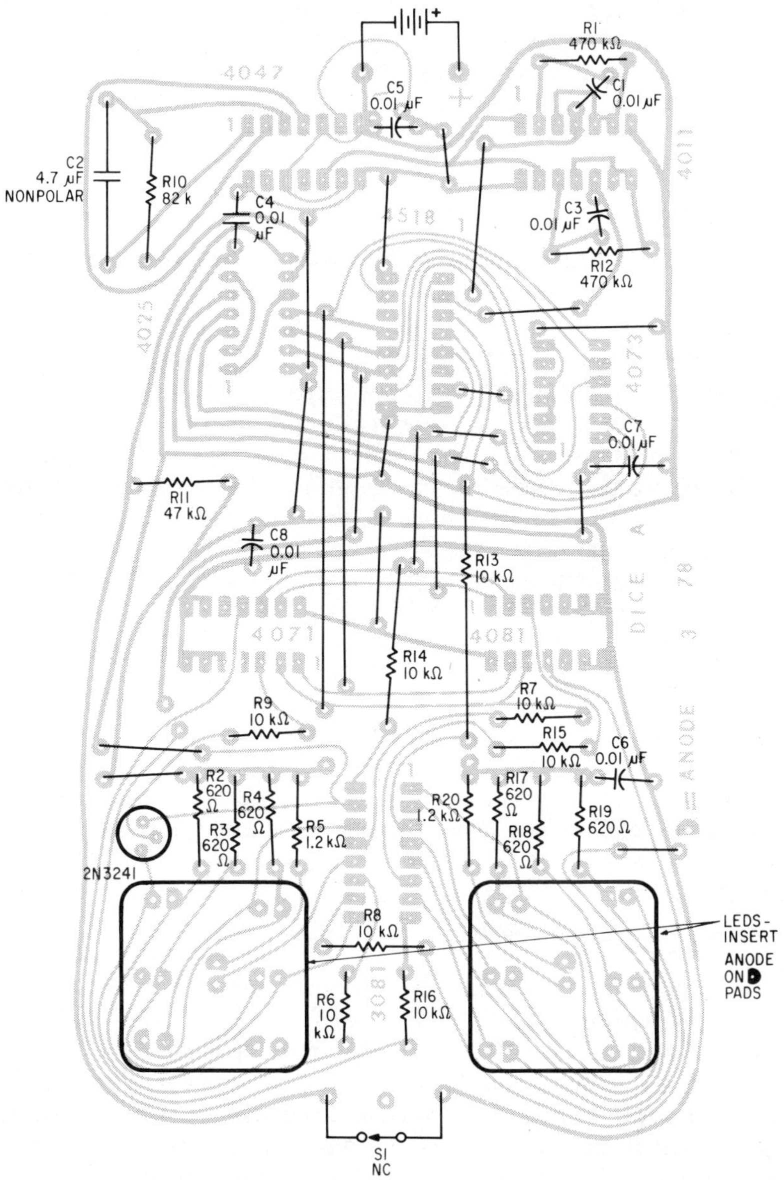

Fig. 5-3 Electronic dice component location diagram.

To prevent the counter from coming out with a number greater than 6, the BCD code for 7 is used to reset the counter to zero. Obviously, zero is not a dice number. To prevent a "throw" ending with a zero on either die, the CD4025 is wired to automatically retrigger the timer if the call is zero. The switch connected to this IC is the "roll-em" trigger for each throw. Sometimes a zero (all LEDs out) will appear, but the timer will automatically reroll the dice. This oddity can be avoided by using the binary equivalents of 0 to 5 instead of 1 to 6 that were used, but the "BCD-to-dice" decoder will become more elaborate. Besides, this adds to the novelty of the game.

Construction

As with most of the circuits in this book, the printed circuit board has no hidden construction quirks. Component location is given in Fig. 5-3 on the previous page. Adequate by-passing is provided to prevent glitches. The "roll-em" switch-S1 must be a momentary-open type to start the timer.

The power supply may provide anything from 5 to 15 volts. Battery operation at 5 volts is feasible. The greater the supply voltage, the brighter the lights but also the greater the current drain, making battery operation less practical. The major current goes to light the LEDs. The CMOS devices use only milliwatts of power.

With this electronic dice game, your friends will beat a path to your door but please, don't let the stakes go above $1000 a throw!

Digital Roulette

For those who are real gamblers, there is nothing like a spin of the wheel to arouse excitement. After the fun of the electronic dice has worn thin, follow up with roulette (Fig. 5-4). The novelty of a digital read-out will add to the fun.

Circuit Description

The circuit (Fig. 5-5) uses CMOS digital logic to provide an equivalent of the big wheel.

An oscillator, the CA3094, is used to provide a series of pulses to a pair of divide-by-10 counters. As long as oscillations continue, the counters, through their associated drivers, change the digits at a rapid rate. This particular oscillator circuit is self-quenching. That is, after a measured number of seconds, oscillation will stop. What is most interesting is that the oscillator keeps slowing down. As a result, the rate of change of numbers on the read-out slows down. This gives the effect of the roulette wheel slowly coming to a stop. The CA3094 is operating in a voltage-controlled oscillator (VCO) circuit. The voltage at pin 5 and the value of C1 determine the frequency of oscillation. Capacitor C2 is charged to the supply voltage through the start switch S1. When S1 is released, the capacitor slowly discharges, thereby lowering the voltage of pin 5 which

Fig. 5-4 Digital roulette finished assembly.

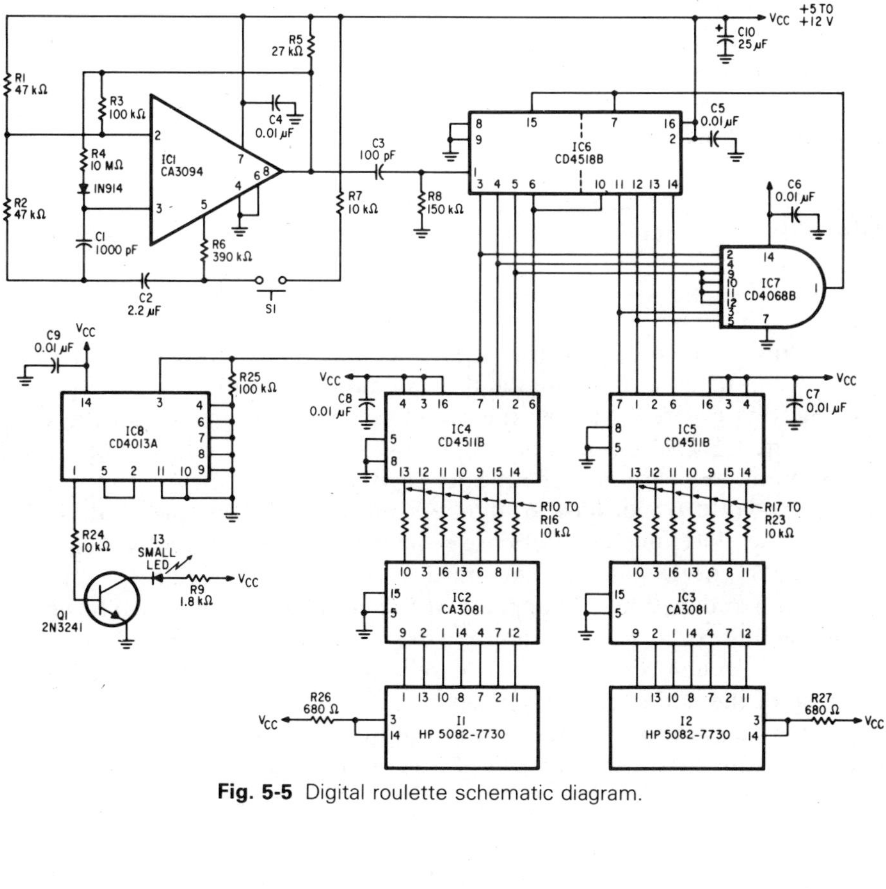

Fig. 5-5 Digital roulette schematic diagram.

Parts List

C1—1000-pF, 100-volt ceramic
C2—2.2-μF, 25-volt electrolytic
C3—100-pF, 100-volt ceramic
C4 through C9—0.01-μF, 100-volt ceramic
C10—25-μF, 25-volt electrolytic
D1—1N914 diode
I1, I2—HP 5082-7730 display
I3—Any miniature LED
IC1—RCA CA3094 power op amp (or SK-3617)
IC2, IC3—RCA CA3081 high-current npn array (or SK-3550)
IC4, IC5—4511A BCD-to-7-segment latch/decoder/driver
IC6—4518B dual synchronous BCD divide-by-10 counter
IC7—4068B 8-input NAND gate
IC8—4013A dual D flip-flop (or SK-4013)
Q1—2N3241 small-signal npn transistor
R1, R2—47 kilohms
R3, R25—100 kilohms
R4—10 megohms
R5—27 kilohms
R6—390 kilohms
R7, R10 through R24—10 kilohms
R8—150 kilohms
R9—1.8 kilohms
R26, R27—680 ohms
S1—SPST normally open momentary switch
Misc.—Printed circuit board, suitable enclosure, IC sockets, terminals, etc.

---

in turn lowers the oscillator frequency. At some point, the capacitor voltage will fall below a minimum and oscillation will stop.

At this point, whatever the counters read will appear on the two digits. The CD4511 devices are BCD-to-7-segment decoders. They convert the counter output to the proper digital display logic.

A roulette wheel has 36 numbers plus double zero ØØ, and single zero Ø. The electronic counters are reset after a count of 36 by means of the CD4068 AND gate. Its inputs are wired to recognize the binary values of 3 and 7. When 37 appears at the counter output, the AND gate responds by resetting the counters to zero. The counters, therefore, read from ØØ to 36 and reset. The single zero of the wheel has been omitted because of the added circuitry needed.

One more item is necessary. Is the number red or black? This display is provided by a single LED. If it is on, the call is red; if it is off, the

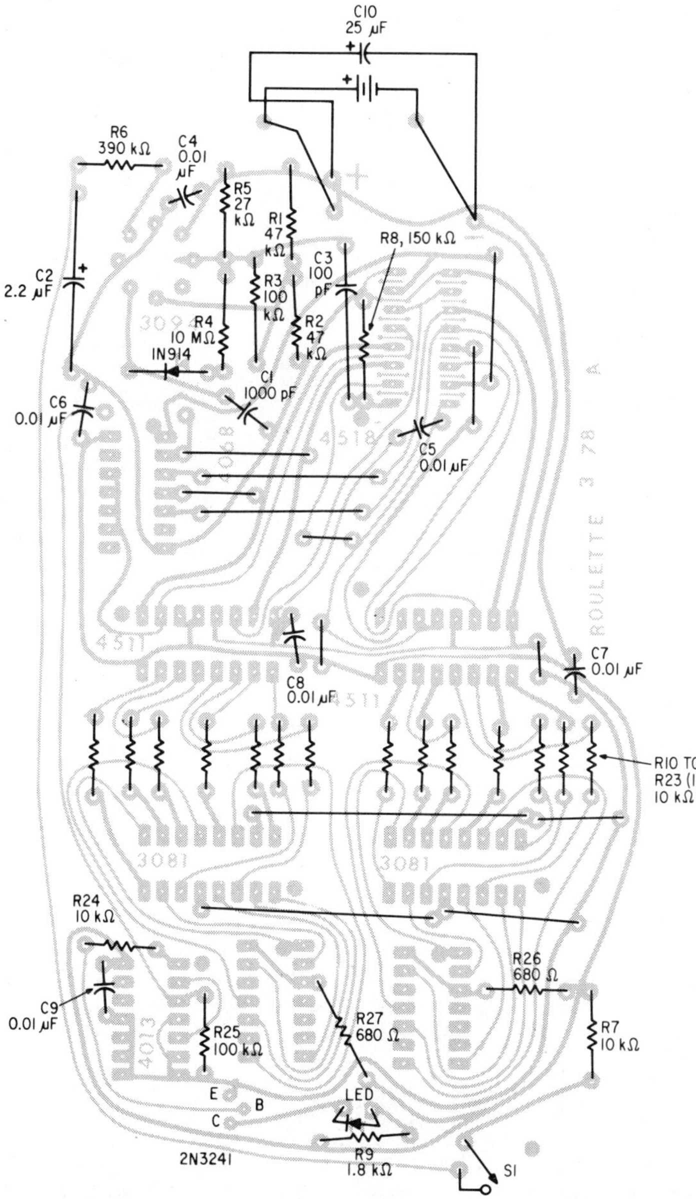

Fig. 5-6 Digital roulette component location diagram.

call is black. Switching between red and black is done by the CD4013 flip-flop. The flip-flop rate of switching is different than the reset cycle. This allows red to appear on either odd or even numbers. Betting can be on the number, odd or even, or red or black.

Construction

Again, there is nothing critical with parts location, shown on the previous page in Fig. 5-6. It is good practice to by-pass the V_{cc} line with a 0.01 μF capacitor at or near each IC. This circuit will work equally well if it is hand wired.

The switch S1 must be a normally open, momentary close type so capacitor C2 can discharge through the oscillator circuit.

Any supply voltage from 5 to 15 volts may be used. As with the dice game, battery operation is possible. It would be advisable to increase the values of R26 and R27 to reduce battery drain through the digital readout, though there will be some sacrifice in brightness. Have fun spinning the big wheel!

Appendix

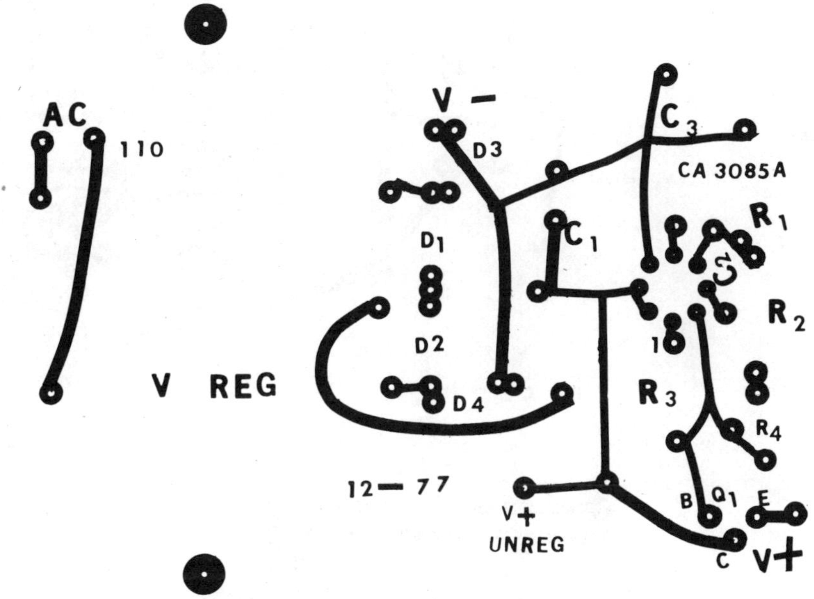

Fig. A-1 Medium current power supply.

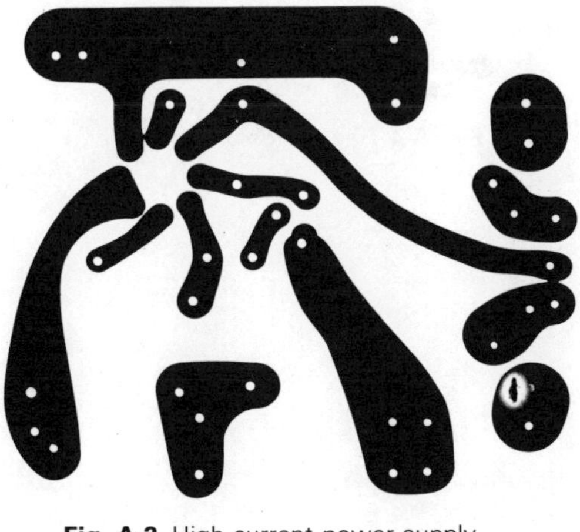

Fig. A-2 High current power supply.

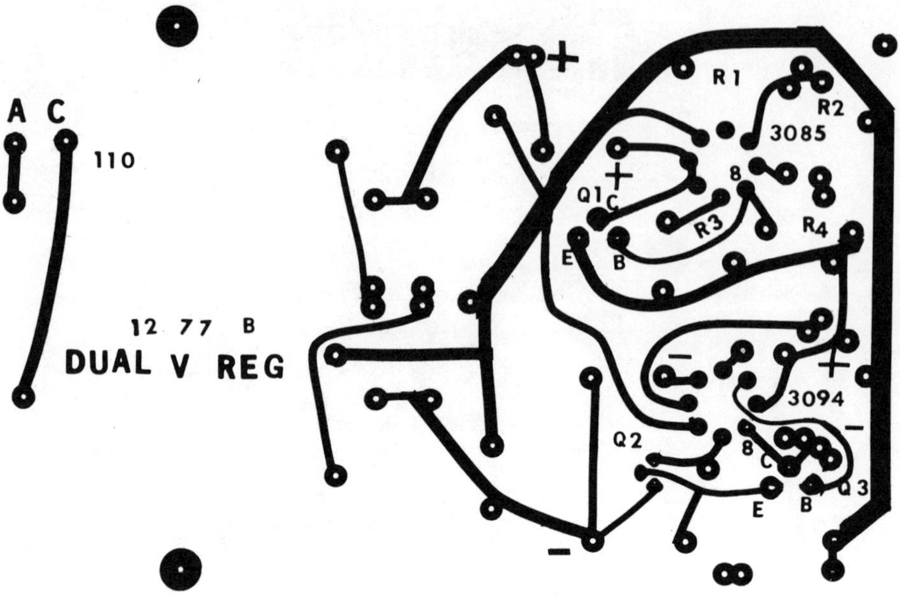

Fig. A-3 Tracking dual supply.

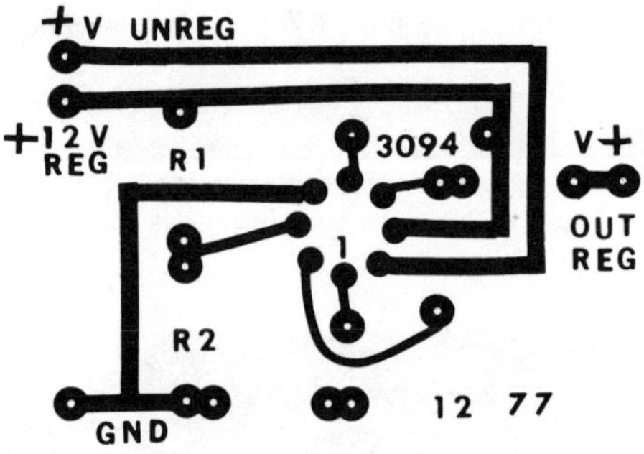

Fig. A-4 Regulated voltage divider.

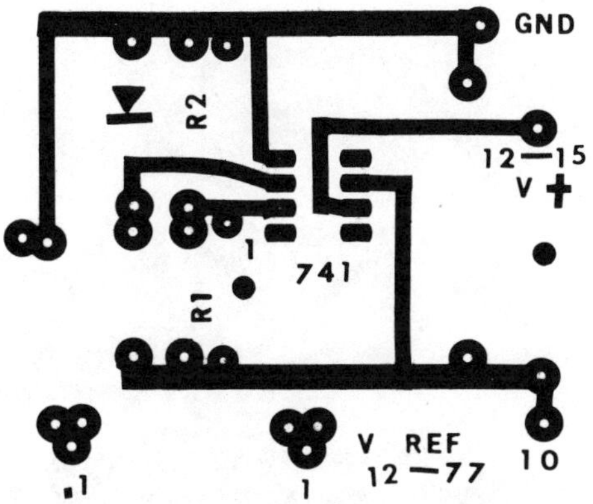

Fig. A-5 Stable voltage reference.

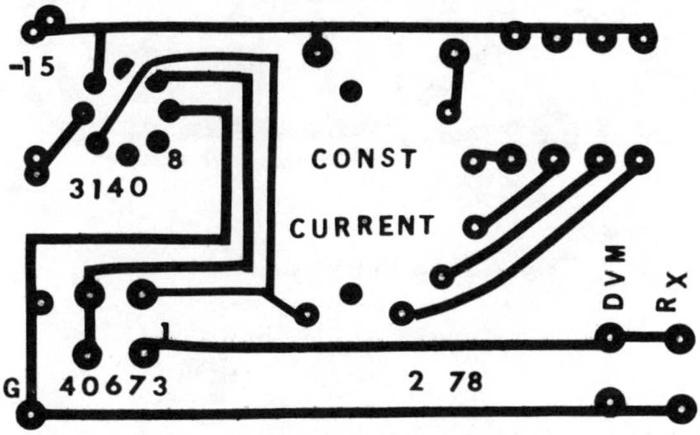

Fig. A-6 Constant current source.

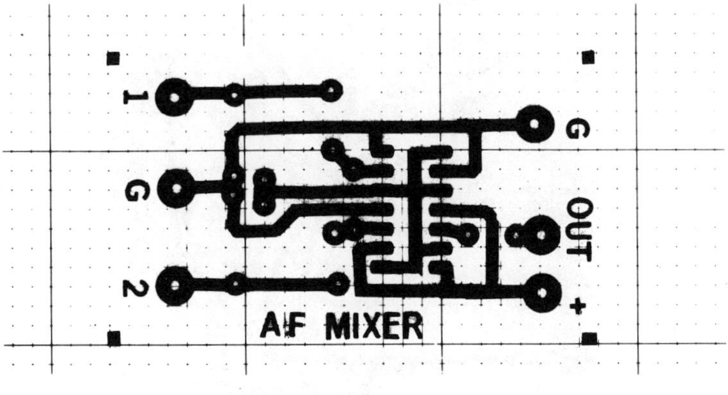

Fig. A-7 Audio mixer.

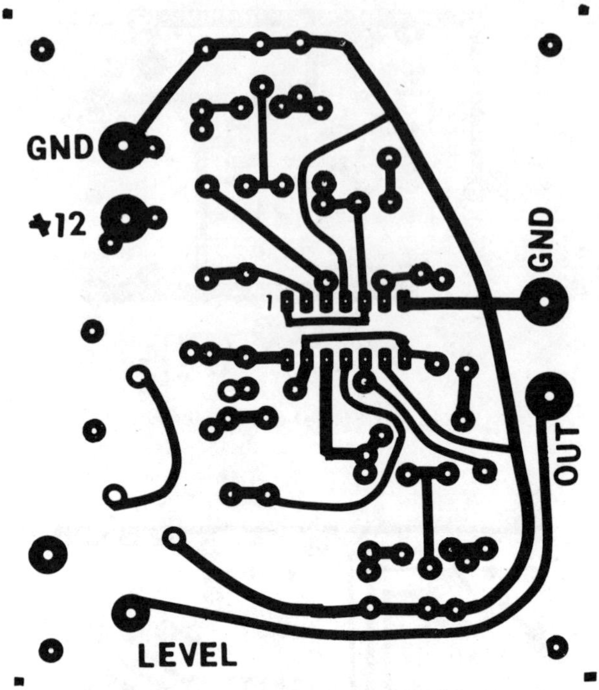

Fig. A-8 Two-tone audio oscillator.

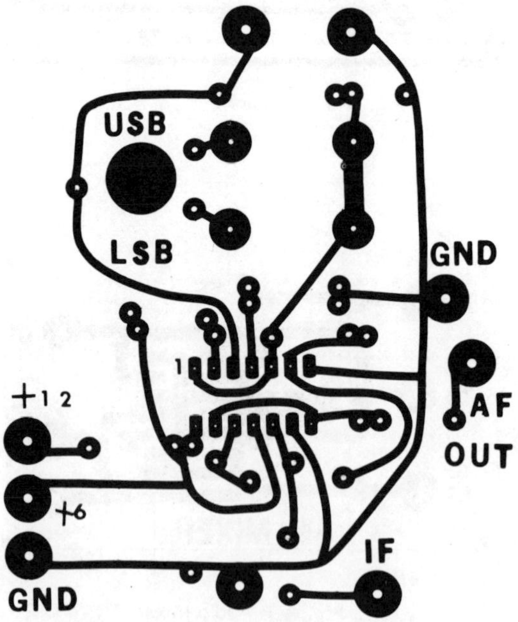

Fig. A-9 SSB detector.

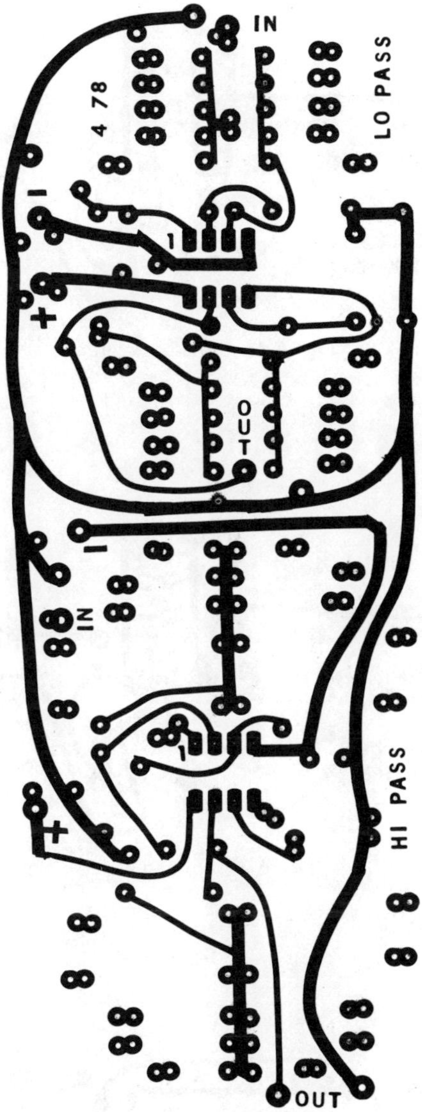

Fig. A-10 High- and low-pass filter.

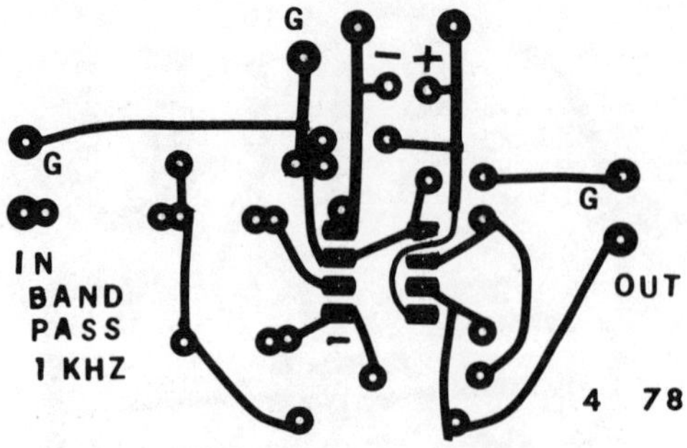

Fig. A-11 Band-pass filter.

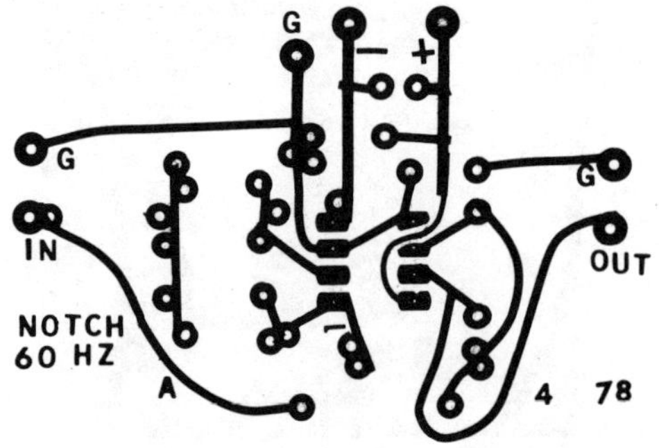

Fig. A-12 Notch filter.

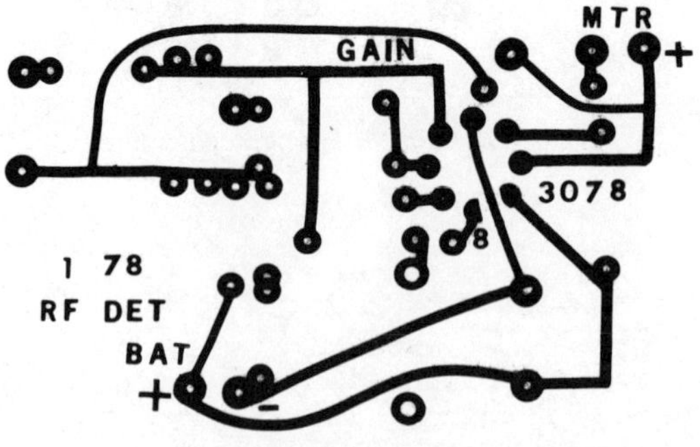

Fig. A-13 RF detector.

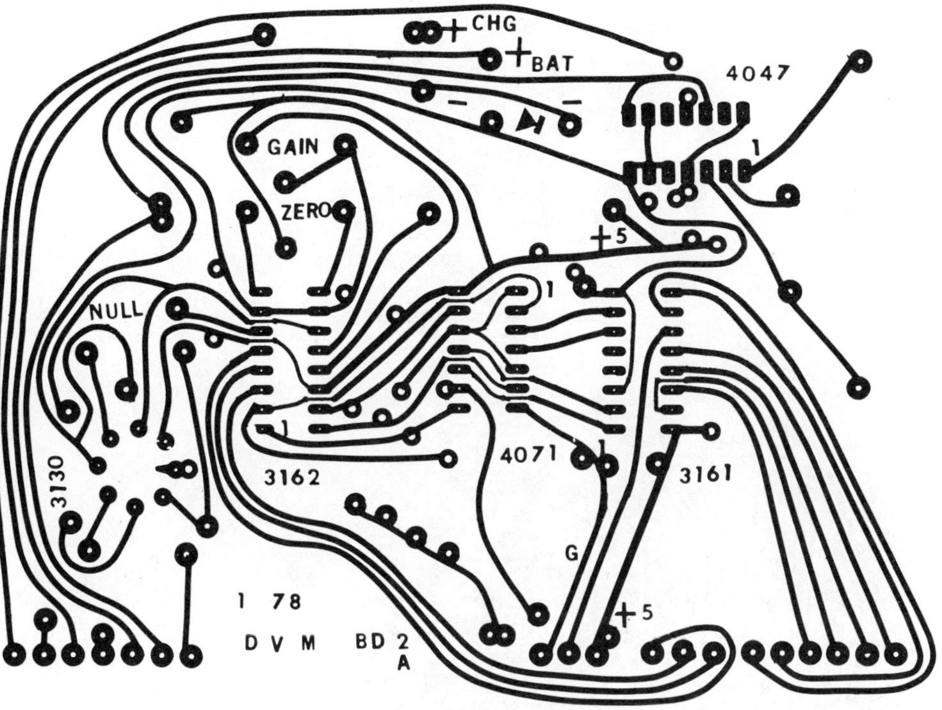

Fig. A-14 DVM horizontal board.

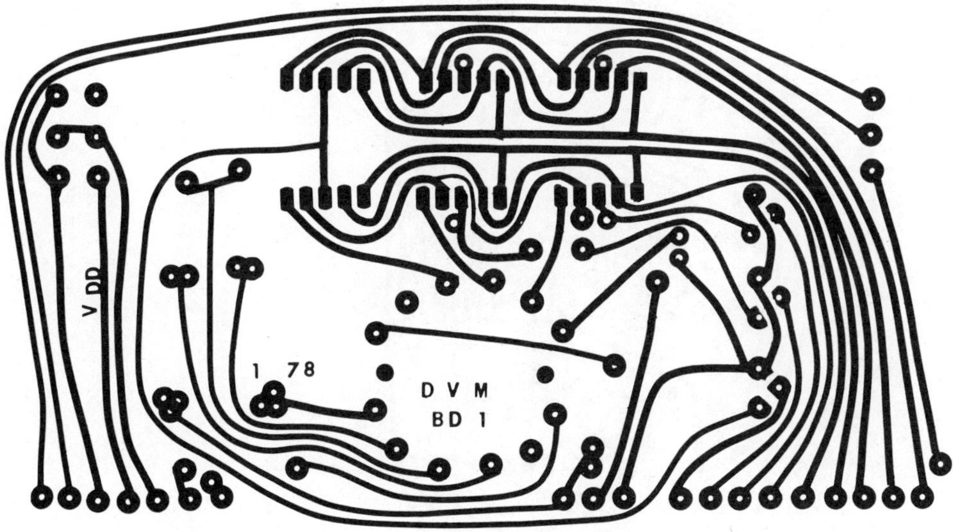

Fig. A-15 DVM vertical board.

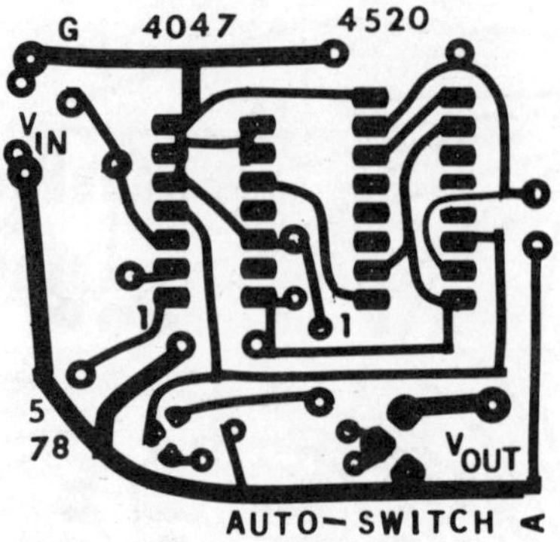

Fig. A-16 Battery saver.

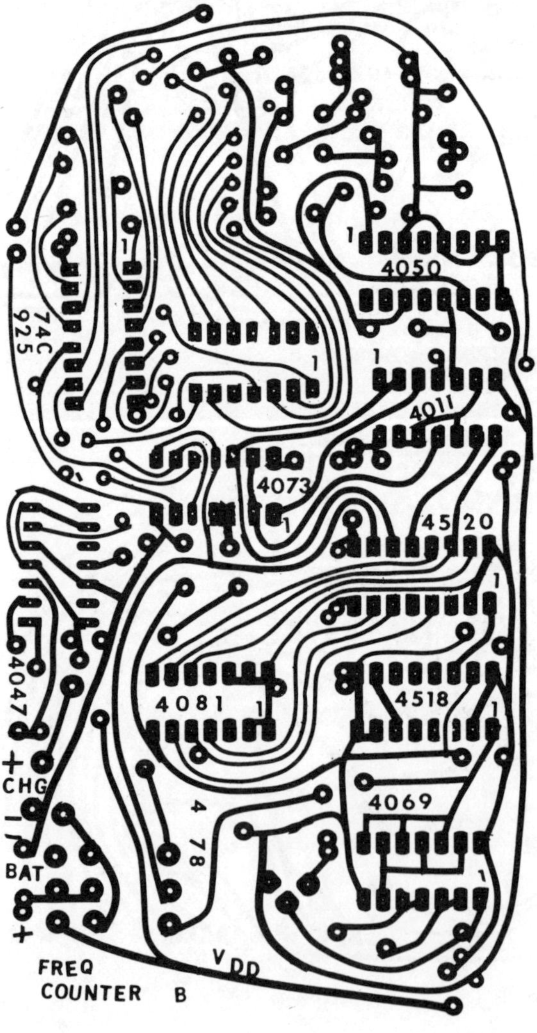

Fig. A-17 Frequency counter.

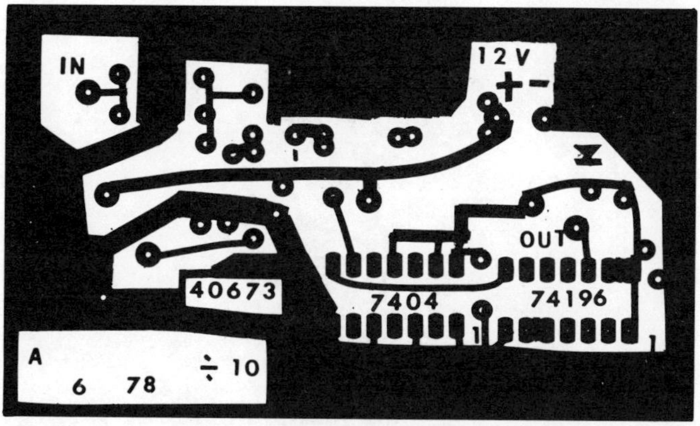

Fig. A-18 Divide-by-10 pre-scaler.

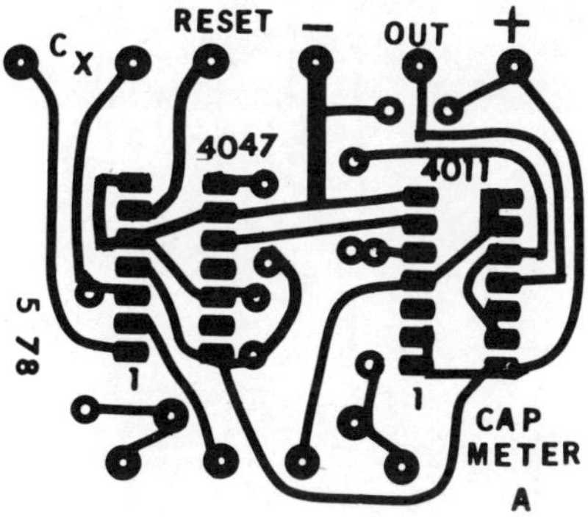

Fig. A-19 Capacitance meter.

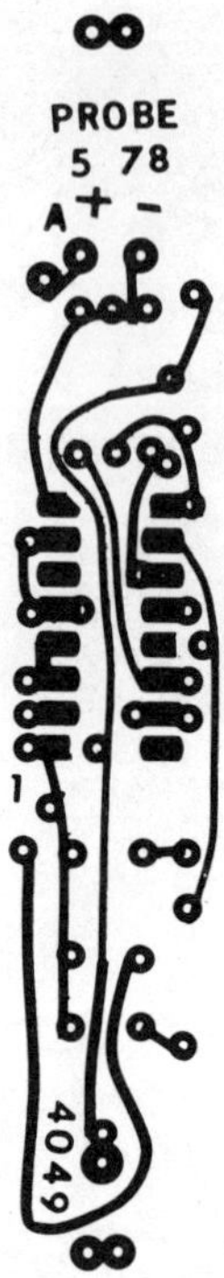

Fig. A-20 Digital probe.

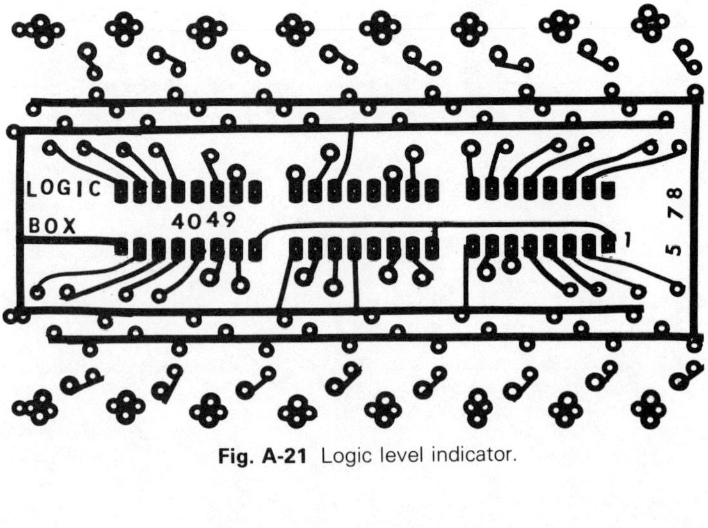

Fig. A-21 Logic level indicator.

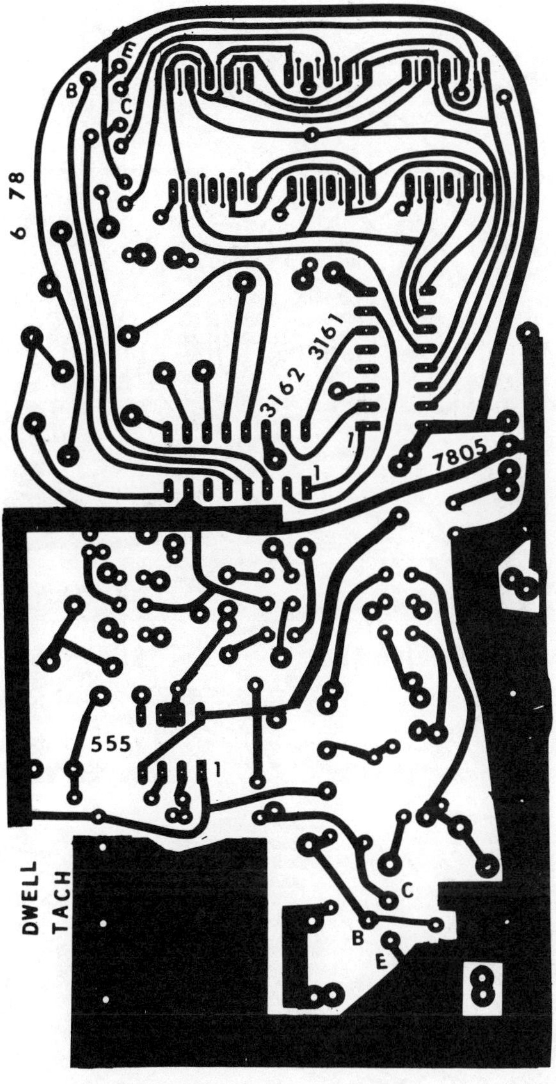

Fig. A-22 Dwell-tach.

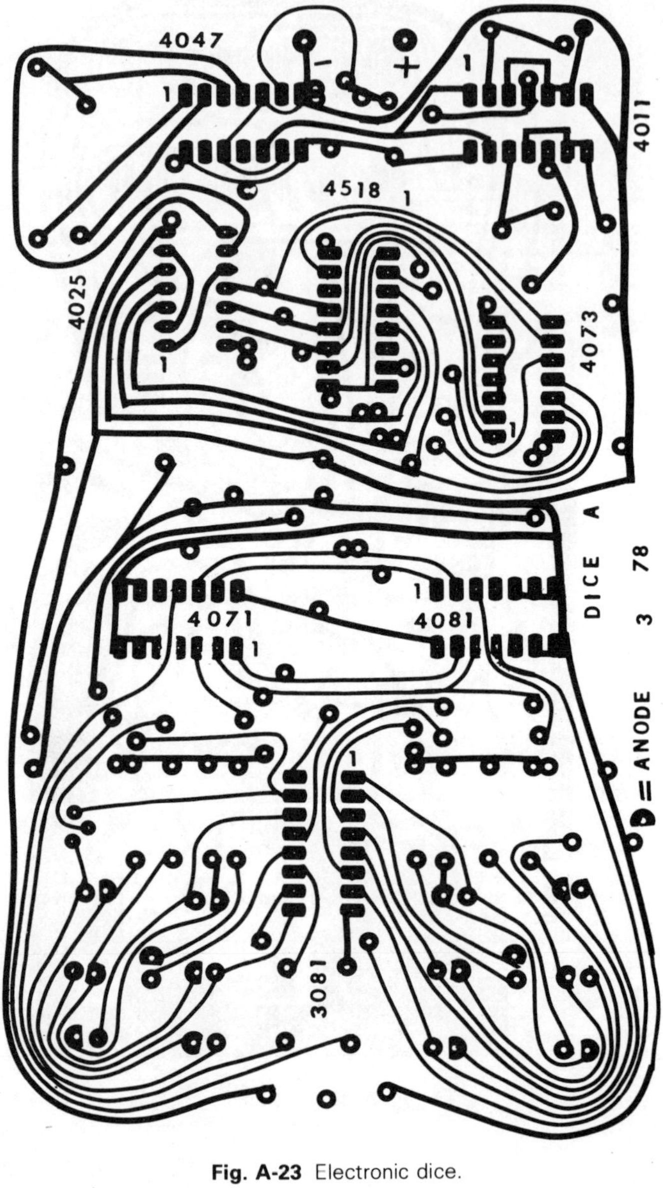

Fig. A-23 Electronic dice.

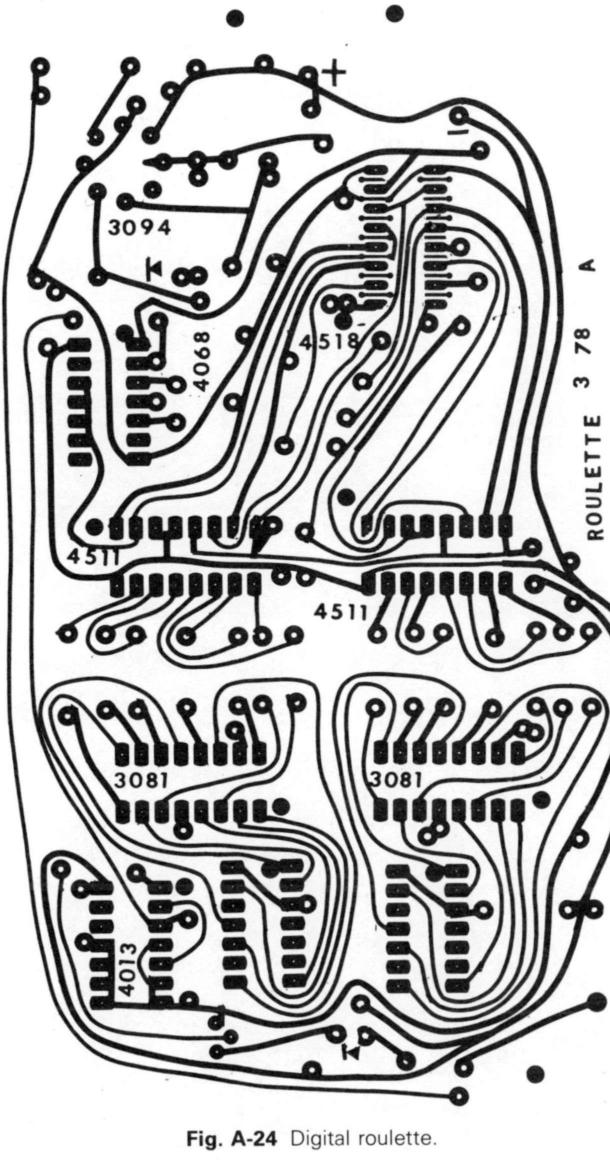

Fig. A-24 Digital roulette.